Queensland Department of Primary Industries
Information Series QI88018

FRUIT AND NUT CROPS

a disease management guide

Edited by

Denis M. Persley
Plant Pathology Branch

Ken G. Pegg
Plant Pathology Branch

Jim R. Syme
Division of Plant Industry

Coordinated by Plant Pathology Branch

Department of Primary Industries
Queensland Government
Brisbane 1989

ISSN 0727-6273

AGDEX 202/633

First published in 1989

National Library of Australia Cataloguing-in-Publication Data

Fruit and nut crops: a disease management guide

ISBN 0 7242 2555 2

1. Fruit – Diseases and pests – Control. 2. Nuts – Diseases and pests – Control.
I. Persley, Denis M. II. Pegg, K. (Kenneth), 1937– . III. Syme, Jim R.
IV. Queensland. Plant Pathology Branch. V. Queensland. Dept. of Primary Industries.
(Series: Queensland Department of Primary Industries information series; QI88018).

634'.049

Queensland Department of Primary Industries
GPO Box 46
Brisbane Qld 4001

Published by direction
of
the Hon. N. J. Harper, M.L.A.
Minister for Primary Industries

Foreword

This book provides descriptions and management recommendations for diseases of fruit and nut crops in Queensland. It is the first in a series which will replace the current *Handbook of plant diseases in colour - Volume 1 (second edition)*. A third edition of the handbook is not published in its original format because it has not proved feasible to update sections, and because chemical registrations and disease management recommendations are constantly changing. Therefore the original handbook has been separated into two parts, with current recommendations being published separately from the colour sections.

The future in plant pathology lies in an integrated approach to disease control where resistant cultivars, crop rotation, biological control and other cultural practices, as well as fungicides, are all important in our bid to economically reduce disease and produce quality fruit and nuts. These measures are emphasised in this publication, with many being specifically designed to counter the problem of resistance to fungicides and to reduce the risk of pesticide residues in fruit.

Recommendations for fungicide use are correct at the time of printing but changes in registrations occur regularly. If in doubt concerning the use of a particular chemical on a crop, consult the extension horticulturalists of this Department who have access to the computer data base 'Infopest'. This system is continually updated and provides information on all current registrations of agricultural chemicals available in Queensland. New information on management practices is also available through 'Infopest'.

I wish to compliment the work of Mr D. M. Persley, Mr K. G. Pegg and Dr J. R. Syme in editing this volume. I also gratefully acknowledge the assistance from staff of Standards and Information Branches in providing data and organising the publication of this volume.

I. F. Muirhead
Director
Plant Pathology Branch

Contents

Acknowledgements

The following members of Plant Pathology Branch made significant contributions to the text:

S. R. Dullahide, R. S. Greber, J. B. Heaton, D. G. Hutton, G. I. Johnson, M. L. Moffett, P. C. O'Brien, K. G. Pegg, and G. R. Stirling.

Miss C. L. Wright provided valuable secretarial assistance. A. Nikulin and J. Rochecouste assisted with the initial editing of text on Infopest.

Ms R. Lancaster, Information Branch, assisted in the editing and production of the book.

Glossary

Anthracnose - a disease usually characterised by depressed spots lined with pink spore masses during moist conditions. Characteristic of infection by the fungus Colletotrichum.

Ascospore - a fungal spore formed during the sexual stage of a life cycle. An ascospore is carried in a sac-like structure called an ascus. Ascospores often form the initial disease inoculum at the beginning of a season, for example spring infection by apple black spot disease.

Button - star-shaped green calyx where the stem meets the fruit in citrus.

Calyx (calyces) - outermost part of a flower, consisting usually of green leaf-like structures.

Cambium - thin layer of actively dividing cells lying between the xylem (water conducting) and phloem (food conducting) systems of most vascular plants. The cambium is responsible for secondary growth which results in growth in diameter.

Canker - a plant disease in which callusing occurs, killing localised areas of bark on branches or stems.

Chlorosis - a yellowing or absence of normal green colour from plant parts.

Conidium (pl. conidia) - fungal spore produced during the non-sexual stage of the life cycle. Allows fungal dispersal, often with aid of wind or water.

Gummosis - a plant disease where the production of gum from infected plant parts is a distinguishing feature.

Gumming - the production of thick syrup-like gum in response to infection by a pathogen.

Greenlife - fresh storage life of fruit.

Lenticel - a minute hole or pore in the stem or fruit allowing gas exchange. Often a point of entry for bacteria.

Melanose - a blemish of citrus fruit caused by a fungus. It develops as round, gummy lumps or spots which may be arranged on the fruit in various patterns such as circles or as tear streaks or stains. If the spots are so dense as to cause a crust, this scar tissue later cracks to give a 'mudcake' effect.

Mummies - dry shrivelled hard fruit. For example, stone fruit affected by brown rot mummify if left on the tree or ground.

Oospore - a thick walled fungal resting spore which is resistant to adverse conditions. Produced by the water mould fungi, for example Pythium and Phytophthora.

Pathogen - a parasite able to cause disease in a host plant. The major groups of plant pathogens are fungi, bacteria, nematodes and viruses.

Perithecium (pl. perithecia) - flask-shaped spore-bearing structure in which ascospores develop.

Sclerotia - small hard dark fungal structures. They can remain dormant in the soil for long periods and germinate upon the return of favourable conditions. They are an important means of survival for some fungal pathogens, for example *Sclerotium rolfsii.*

Shoestrings - black string-like threads produced by some fungi, for example, the root pathogen *Armillaria luteobubalina*, commonly known as the 'shoestring fungus'.

Spore - a minute propagative unit of a fungus which functions as a seed.

Stolons - a long, slender, modified stem which produced new plants. For example, stolons or 'runners' arise from the crown of strawberry plants.

Apples

FIELD DISEASES

Fungi

Armillaria root rot
Armillaria luteobubalina

Symptoms

A decline and death of trees, generally in groups. A cream-coloured fungal growth, sometimes in fans, is seen just beneath the bark of the crown and large roots. The underlying wood will be discoloured and a strong mushroom smell is often detected. Black threads of the fungus, known as shoestrings, may be visible on the surface of the roots. In wet weather, small honey-coloured mushrooms with widely separated gills occasionally form at the base of an affected tree.

Source of infection and spread

The fungus may survive in the soil on stumps and roots of various trees for many years. It grows through the soil, invading living roots it contacts. Once established in a tree, it may spread to other trees by root contact.

Importance

May cause significant losses of mature trees.

Control

Remove diseased trees and as many roots as practicable. Do not replant for at least two years.

Replanting infested sites

First remove tree roots and cultivate to a depth of 300 mm. Treat a 3.6 m^2 at each site with methyl bromide (two 680 g cans) under a 4 m^2 plastic sheet with edges imbedded to a depth of 150 mm. Leave the sheet in place for two days.

Bitter rot
Glomerella cingulata

Symptoms

The disease affects fruit only. Small circular light brown slightly sunken spots enlarging rapidly to rot large areas of the fruit. Raised pink spore masses, often in concentric rings, appear near the centre of each spot. The spots generally appear when the fruit approach maturity and are soft but not distinctly mushy or leathery.

Source of infection and spread

The fungus overwinters in fruit mummies from the previous season and in bark crevices. Spores are released from these areas in warm humid weather to infect fruit. The pink spore masses which then develop on fruit provide large numbers of spores allowing rapid spread of the disease.

As ripe fruit is most susceptible, the disease seldom appears until late in the season.

Importance

Summer disease which may be serious as fruit nears maturity. It can be controlled by spraying and attention to orchard hygiene.

Control

1. Remove and destroy all mummified fruit at pruning and diseased fruit during the season.
2. Spray with fungicides.

Stage: Late season (late November to maturity)

Chemicals: mancozeb, zineb, maneb + zineb, ziram.

Remarks:

1. Spray at 21-day intervals with above fungicides from November to near harvest.
2. Remove and bury fruit affected by bitter rot.

Black rot
Botryosphaeria ribis

Symptoms

Affects wood, leaves and fruit.

Branches: small elliptical areas develop; the bark shrinks, lightens in colour, and then becomes sunken and separated from healthy bark by a crevice. The fungus grows radially so that in old cankers there may be a number of crevices arranged concentrically. Eventually, the branch becomes girdled and dies.

Leaves: greyish-brown spots with darker surrounding zones giving a frog-eye appearance.

Fruit: small brown circular spots rapidly increasing in size to involve the whole fruit. The colour becomes jet black and the fruit shrivels to a mummified condition.

Importance

The disease is uncommon in Queensland.

Control

Remove affected branches during pruning and destroy mummified fruit.

Black spot
Venturia inaequalis

Symptoms

Leaves: light olive green spots, about 3 mm in diameter with a diffuse, irregular velvety growth, mostly on the undersurface of young leaves. Spots darken with age, eventually becoming black and may coalesce to cover most of the leaf. Leaves are small, distorted and may fall.

Fruit: small dark spots which enlarge, becoming brown and corky with a black broken margin.

Source of infection and spread

During winter, the fungus grows within old diseased apple leaves beneath the trees. The fungus forms minute fruiting bodies (perithecia) in which large numbers of primary spores (ascospores) are produced. With the onset of spring rains, these are forcibly ejected into the air and carried by wind and rain to developing apple leaves and flowers where infection occurs if conditions are suitable. By midsummer, almost all primary spores have been released. Two weeks after infection, spots appear on the leaves and fruit. On the surface of these spots, secondary spores (conidia) are produced and are spread to adjacent trees by wind and rain. Showery weather favours new infections. Secondary spores are produced until autumn leaf fall.

Importance

Can be serious, particularly in wet seasons, but can be controlled by a regular spray programme.

Control

1. Spray with recommended fungicides, especially from green tip to 95% petal fall.
2. Apply the recommended pre-leaf fall spray.
3. During winter disc diseased leaves and fruit into the soil to reduce disease carry over to the next season.

Stage: green tip.

Chemicals: copper oxychloride or copper hydroxide or dithianon.

Alternative fungicides are mancozeb, thiram, ziram or zineb.

1. Thorough coverage with fungicide required.
2. If using an integrated mite control programme for pest mites avoid using mancozeb or metiram.
3. Copper sprays may cause fruit russet if green tip is variable, especially on cv. Delicious.

Stage: blossoming (blossom break to petal fall).

Chemicals: protectant fungicides – mancozeb, metiram, thiram or ziram.

Eradicant fungicides – bitertanol + Agridex®, dodine, fenarimol, or triforine.

1. Apply protectant fungicides every 10 to 14 days, except during dry conditions when the interval may be extended to two to three weeks.
2. If rain prevents spraying apply an eradicant fungicide according to black spot warnings. These fungicides eradicate infections if applied between 48 to 120 hours after the start of a primary ascospore infection period. Apply dodine within two days of warning, or triforine within three days of warning, or fenarimol within four days of warning or bitertanol + Agridex® within five days of warning.
3. Include mancozeb or ziram with fenarimol if prolonged wet weather occurs.
4. Do not use fenarimol on cv. Delicious from pink bud to full bloom.
5. Do not use triforine on cvs. Golden Delicious or Cox's Orange Pippin.
6. Do not use dodine under cold slow-drying conditions as fruit russet may occur.

Stage: mid season (from petal fall to December).

Chemicals: as for blossoming stage.

1. Spray at 14 to 21 day intervals.
2. As for blossoming stage.

Stage: late stage: (December to maturity).

Chemicals: mancozeb or zineb or ziram.

1. At least two sprays with one of these fungicides are required before harvest.
2. Apply at 21-day intervals. These sprays will also control sooty blotch and flyspeck.
3. Fruit affected by bitter rot should be removed and buried.

Note: Use eradicant sprays alternately as advised by the black spot warning service. This minimises the development of fungicide resistance. Benomyl, carbendazim and thiophanate-methyl are not recommended due to resistance by the black spot fungus.

Stage: post-harvest.

Apply urea at 5 kg/100 L. Spray trees thoroughly from 50% leaf fall onwards and spray leaves beneath trees. This reduces carry over of the disease.

Brown rot

Sclerotinia fructicola

Symptoms

Fruit only are affected. Spots are at first brown and circular, often with black areas near the centre. The whole fruit eventually decays and in warm moist weather, tufts of spores develop on the surface. Rotted fruit is soft but not mushy.

Source of infection and spread

Brown rot is principally a post-harvest disease but may occur on trees near diseased stone fruits. Only damaged mature fruit is susceptible.

Importance

Rare on apples.

Control

1. Handle fruit carefully to avoid damaging the skin.
2. Keep packing sheds free from rotting fruit.
3. Cool fruit promptly to restrict disease development.

Collar rot

Sclerotium rolfsii

Symptoms

Leaves of affected trees turn reddish or greyish purple in colour and wilt, eventually becoming brown and papery. Examination of the trunk at or just below ground level shows a rot of the bark and wood often completely encircling the stem. White wefts of fungal growth may cover the affected area when the soil is moist. The most characteristic symptom is the production of small spherical, resting bodies (sclerotia) on the affected area or in the nearby soil.

Source of infection and spread

The fungus is a common soil inhabitant which attacks a large number of crop plants and weeds. It is often serious in apple trees planted in areas previously used for vegetables.

Importance

Generally a problem only where apples are being planted in old vegetable land. Can be prevented by drenching the soil with a fungicide at planting.

Control

1. When planting apples in old vegetable land, drench the base of young trees with the recommended fungicide at planting and again in midsummer.
2. Inspect trees regularly and, if any show symptoms, remove the soil from the crown, cut away the diseased wood and treat the area with the recommended wound dressing. Leave the affected area exposed for at least 10 weeks. This treatment is effective only when the disease is detected early.

Fungicide drench

For control of collar rot when nursery trees are being planted in old vegetable land or infested soil; prepare a suspension containing 10 g/L of quintozene (750 g/kg). Drench the butt of the tree with 0.5 to 1 L of the mixture.

Wound dressing

After diseased wood has been removed, paint with a thick slurry of copper oxychloride. Leave the painted area exposed for at least three months before covering with soil.

Dieback

Wood rotting bracket fungi

Symptoms

At first, a browning and 'papering' of the outer bark of affected branches. These rapidly lose vigour and die. Fungal brackets (fruiting bodies) develop on the diseased wood. Some characteristics of the brackets are given below.

Trametes versicolor. Up to 30 mm across, with various coloured bands, and a smooth surface on the upper side and a honeycomb appearance underneath. The most common wood-rotting fungus affecting apple trees in Queensland.

Trametes velutina. Similar to *Trametes versicolor* except that the upper side is hairy and the bands are grey to brown.

Pycnoporus coccineus. Up to 70 mm across, bright orange-red in colour with a honeycomb appearance underneath. A weaker pathogen but often infects newly strap-grafted branches.

Source of infection and spread

The brackets or fruiting bodies produced on old branches and stumps of infected apple trees contain large numbers of spores. These are released and carried by wind to nearby apple trees where germination occurs if conditions are suitable. Entry is usually gained through damaged tissues such as unprotected pruning cuts and natural wounds. Once established, the fungus grows down through the heartwood causing dieback as it progresses. The condition is more common in mature trees which have suffered stress from drought, hail, sunscald, mechanical root damage, overcropping, poor nutrition or severe winter pruning.

Importance

Generally serious only where trees have been badly stressed by severe pruning, drought or deep cultivation.

Control

1. Prune trees carefully.
2. Treat cuts with the recommended wound dressing to promote rapid healing.

Wound dressing

After pruning out affected branches at least 200 mm below any visible discolouration, paint the cut surface immediately with anti-mould household paint. Wash the pruning saw blade with methylated spirits after each cut. With healthy trees, prune **minimally** to shape trees. When large pruning cuts are necessary, they should be done in early summer rather than winter.

Fly speck

Schizothyrium pomi

Symptoms

On fruit, roughly circular groups of small black dots resembling marks made by flies on walls. The dots are superficial and easily rubbed off. It is often seen in association with smoky blotch.

General

The disease is a problem on ripening fruit and is favoured by cool wet weather.

Importance

Late summer disease of minor importance in well-sprayed orchards.

Control

Spray with recommended fungicides.

Stage: late season (December to maturity).

Chemicals: mancozeb, zineb, maneb + zineb, ziram.

1. At least two sprays with this fungicide are required before harvest.
2. Spray at 21-day intervals.
3. Fruit affected by bitter rot should be removed and buried.
4. The recommended spray programme for black spot will also control this disease.

Powdery mildew
Podosphaera leucotricha

Symptoms

Leaves: white to light grey powdery patches which develop rapidly to envelop both leaf surfaces causing curling, stunting and death.

Twigs: a similar white, powdery growth to that on leaves causing stunting and dieback. Tips of twigs are extremely vulnerable and fruit and leaf buds may become coated with mildew and fail to develop fully. Affected blossoms frequently wither and fruit fails to set.

Fruit: a skin russet composed of a maze of fine lines, often appearing as a solid patch. Fruit become deformed and cracked.

Source of infection and spread

The fungus overwinters in dormant flower and leaf buds. It resumes growth in spring and infects new leaves and blossoms particularly from pink stage onwards. Large numbers of spores are produced and are spread by wind and rain to adjacent trees.

Infection is favoured by high relative humidity and temperatures between 19 and 25°C. The risk of infection decreases as summer temperatures rise; mildew spores being short-lived at temperatures above 28°C. Further outbreaks can occur during autumn when favourable cool conditions return.

Importance

A serious disease particularly in susceptible cultivars in dry seasons. A regular spray programme and thorough pruning during dormancy can control the disease.

Control

1. Spray with fungicides.
2. Remove diseased buds and shoots in winter and diseased blossom trusses in spring.

Stage: blossoming (blossom break to petal fall).

Chemicals: bupirimate, fenarimol, penconazole, triadimefon or triforine.

1. Apply one of the above fungicides at blossom break, full bloom and petal fall to cvs. Jonathan, Granny Smith and Gravenstein.
2. These materials are generally compatible with those recommended for black spot. Check labels for compatibilities and restrictions on usage.
3. A number of these fungicides are also recommended for black spot control.

Stage: mid season (from petal fall to December).

Chemicals: as for blossoming stage.

1. Spray at 21-day intervals.
2. As for blossoming stage.

Stage: late season (December to maturity).

Chemicals: as for blossoming stage.

1. Spray at 21-day intervals until mid-January.
2. As for blossoming stage.
3. Fruit affected by bitter rot should be removed and buried.

Ripe fruit spot
Pezicula alba

Symptoms

On fruit, brown pinpoint spots which develop slowly and become black and sunken with creamy white spore masses in moist weather. Often confused with bitter rot.

General

The disease is a problem on ripening fruit only and is favoured by cool wet weather.

Importance

Late summer disease of minor importance in well-sprayed orchards.

Control

Spray with recommended fungicides.

Stage: mid season (Late November to maturity).

Chemicals: mancozeb, zineb + maneb, zineb and ziram. (as for apple bitter rot control p. 1).

1. At least two sprays with this fungicide are required before harvest.
2. Spray at 21-day intervals.
3. Fruit affected by bitter rot should be removed and buried.

Smoky blotch
Gloeodes pomigena

Symptoms

On fruit, dark grey to dull black blotches of indefinite shape and variable size. The blotches are superficial and easily rubbed off. The disease is often associated with fly speck.

Source of infection and spread

The disease affects ripening fruit, particularly cv. Granny Smith during humid weather.

Importance

Late summer disease of minor importance in well-sprayed orchards. The disease causes a downgrading of fruit quality.

Control

Spray with recommended fungicides. The spray programme used for black spot will also control this disease (p. 2).

White root rot
Rosellinia necatrix

Symptoms

Trees develop an unthrifty appearance, leaf-yellowing, cessation of shoot growth, small leaves, premature leaf-fall and small shrivelled fruit.

Crown: bark of the crown roots and the base of the trunk shows a dark wet rot. A distinct sharp margin is evident between healthy and affected bark.

A thin layer of white fungal growth occurs under the bark.

Roots: dark wet surface rot, etching into healthy internal wood. Affected roots are often covered with white strands of fungal growth which are growing into soil and leaf litter.

Source of infection and spread

The fungus survives in the soil on the roots of various trees for many years. Young apple trees planted into infested soils become infected and die.

Importance

Can cause serious tree losses in some orchards.

Control

Remove diseased trees with as many roots as possible and fumigate site with methyl bromide as recommended for control of Armillaria root rot (p. 1).

Nematodes

Root-lesion nematode

Pratylenchus penetrans

Symptoms

Young replanted trees are stunted and unproductive. Root systems of affected trees are small and discoloured. Established trees may also be affected showing reduced growth and twig dieback.

Importance

Can be serious where preplant fumigation is not practised.

Control

1. Fumigate replant sites with the recommended nematicide.
2. Do not replant too soon after removing old trees. Allow the area to revert to weeds or plant a recommended cover crop. Do not plant cowpeas, New Zealand blue lupins or rye as these are good hosts for *Pratylenchus penetrans.*

Fumigation

1. Replanting individual sites in established orchards. First remove tree roots and cultivate to a depth of 300 mm. Treat a 3.6 m^2 at each site with methyl bromide (one 680 g can) under a 4 m^2 plastic sheet with edges embedded to a depth of 150 mm. Leave the sheet in place for two days.
2. Replanting old orchard land. Remove tree roots and spell the area for as long as practicable. Allow the area to revert to weeds or cover crop with Sudan grass or maize in the summer and Saia oats, Priebe prairie grass or Wimmera rye grass in the winter. Do not grow cowpeas, rye, or New Zealand blue lupins. Before replanting, prepare soil by ploughing deeply, ripping and discing. Fumigate with EDB 193 at 110 kg/ha.

Viruses

Gravenstein gnarl

Apple flat limb virus

Symptoms

Slight longitudinal flattenings or depressions in shoots and branches. These depressions develop into deep furrows with affected branches much flattened and twisted. In Queensland, this disease has been recorded only on cv. Gravenstein.

Source of infection and spread

The virus is spread in propagating material.

Importance

Of decreasing importance with the reduction in Gravenstein plantings.

Control

Use only virus-tested propagating material.

Green crinkle

Apple green crinkle virus

Symptoms

Affected fruit are small and malformed, sometimes with small russetted areas in the skin depressions. Water-conducting tissues beneath the depressions tend to be greener than normal. Only fruit on one branch of a tree may be affected. In Queensland, this disease has been recorded only on the cv. Granny Smith.

Source of infection and spread

The virus is spread in propagating material.

Importance

Moderate but decreasing in importance with the wide use of virus-tested propagating material.

Control

Use only virus-tested propagating material.

Mosaic

Apple mosaic virus

Symptoms

Irregular white or yellow blotches often with red necrotic centres on the leaves. Symptoms may be confined to a few leaves or a branch. Although not always obvious, yield is reduced.

Source of infection and spread

The virus is spread in propagating material.

Importance

Widespread but decreasing in importance with the wide use of virus-tested propagating material.

Control

Use only virus-tested propagating material.

Ring spot

Apple ring spot virus

Symptoms

Two types of symptoms occur on fruit. The most obvious is a pattern of dark brown concentric, incomplete circles which develop as the fruit ripens. Irregular patches of rough russet, brown in colour, with narrow bands of dark brown tissues around the margins are also common.

Source of infection and spread

The virus is spread in propagating material. Several symptomless (latent) viruses also commonly occur in apple and may reduce yields.

Importance

Widespread but decreasing in importance with the wide use of virus-tested propagating material.

Control

Use only virus-tested propagating material.

POST-HARVEST

Fungi

Blue mould

Penicillium expansum

Symptoms

Fruit at first shows soft pale brown watery spots with the decayed portions sharply separated from the healthy tissues. Spots enlarge rapidly under favourable conditions and may completely envelop the fruit. Under warm moist conditions, at first white then pale blue and finally blue green powdery clumps develop on the surface of the spots. The fruit eventually becomes a soft watery mass with a characteristic musty smell.

Source of infection and spread

Blue mould is principally a post-harvest disease and is rarely found in the orchard. The most important source of the disease is discarded fruit in and around the packing shed. The fungus generally invades the fruit through wounds in the skin. Once established it develops rapidly under warm moist conditions and produces large numbers of spores. However, in dry cold air, there is little or no surface mould, even when the fruit is completely decayed.

In containers, the fungus may spread from fruit to fruit by contact resulting in 'nests' of mouldy fruit. Also, fruit weakened by long periods of poor storage conditions, may be invaded through the breathing pores in the skin resulting in multiple infections known as 'spot waste'.

Importance

The most serious post-harvest disease of apples but can be controlled by fungicide dips and careful attention to hygiene during handling and storage.

Control

1. Handle fruit carefully to avoid damaging the skin.
2. Keep packing sheds clean and free from rotting fruit. Wash graders regularly.
3. Cool fruit promptly to restrict disease development.
4. Reduce dust in packing shed areas by watering roadways and entry points regularly.

Post-harvest fungicide dip

Chemicals: dip in benomyl + iprodione, or carbendazim + iprodione, or thiophanate methyl + iprodione, or imazalil for 30 seconds.

1. The required amount of fungicide should be made into a slurry before adding to the dipping tank. Do not add the powder directly to the tank.
2. Agitate while dipping to keep the fungicide in suspension.
3. Top up regularly with a suspension of the fungicide equivalent in strength to that of the original dip.
4. The fungicide may be included in diphenylamine anti-scald dips **except** where calcium chloride has been added for bitter pit control.
5. Strains of blue mould resistant to benomyl or carbendazim or thiophanate methyl may be controlled by including iprodione with these fungicides or by using imazalil alone.

Disinfectant for packing and grading equipment

Wash regularly with a hypochlorite-based bleach solution used at household rates, for example 150 mL of 3.5% sodium hypochlorite/L of water. As this corrodes some materials, wash thoroughly with clean water after treatment.

Avocados

FIELD DISEASES

Fungi

Anthracnose
Glomerella cingulata

Symptoms

Small light brown circular spots on fruit. Spots enlarge rapidly and change colour to dark brown or black. Centres of spots may be slightly sunken and, if fruit is kept in a moist atmosphere, pink spore masses will develop on the spots. One or more spots may enlarge to cover the side of a fruit. The rot may penetrate deeply into the flesh in a hemispherical fashion.

Source of infection and spread

Large numbers of spores of the fungus are produced on dead twigs, leaves and fruit. These spread through the plantation during warm showery weather, and are deposited on the fruit surface where penetration occurs. Fruit is susceptible to infection from set to harvest. After initial penetration of green fruit, the fungus stops growing and remains dormant or latent in the tissue until the fruit ripens or is damaged. Resuming active growth, it invades the surrounding tissues. Lesions often develop in damaged immature fruit while still on the tree. Fruit can show symptoms from up to three months before harvest and these are usually shed.

Importance

Anthracnose is serious in wetter growing areas, particularly in more susceptible cultivars. The disease can be controlled by a regular field spray programme, post-harvest fungicide treatment and controlled ripening.

Control

1. Spray with recommended fungicides.
2. Ensure good ventilation and rapid drying of the foliage by pruning lower limbs so that the canopy is at least 500 mm above the ground.
3. Prune out dead twigs and branches before flowering. Remove dead leaves entangled in the tree canopy.
4. Control insect pests such as fruit sucking bugs and caterpillars.
5. Handle fruit carefully to avoid damaging the skin.

Fungicide sprays

Spray with copper oxychloride or cupric hydroxide from flowering to harvest (generally September to mid April) at 28-day intervals. During prolonged wet weather, reduce the interval to 14 days. Use sufficient spray to provide good coverage. No wetting agent is necessary with these sprays.

Cultivars

Fuerte, Nabal, Wurtz and Rincon are more susceptible to anthracnose than other cultivars.

Post-harvest control

1. Spray fruit post-harvest with prochloraz (450 g/L) as an overhead spray at 55 ml/100 L for 30 seconds.
2. Fruit with a short (1 to 3 days) green life (the time between harvest and when ripening commences) should be cooled promptly after harvest to 7°C. Lower temperatures after harvest can cause chilling injury.
3. Temperature is critical to anthracnose development during the ripening stage of the avocado. Once fruit starts to ripen, temperatures of 24°C and above will accelerate development of anthracnose. The first onset of ripening cannot be detected by the eye or feel. Anthracnose development is less extensive at 16 to 18°C. The addition of ethylene will reduce the variation in ripening between fruit. Once avocados have ripened to an easily detectable softening stage, fruit may be held between 2 and 4°C and as low as 2°C for ripe and ready to eat avocados.

Phytophthora root rot
Phytophthora cinnamomi

Symptoms

Leaves: Pale green and wilted, falling readily. Shoots die back from the tips so that eventually the tree is reduced to a bare framework of dying branches. Death of the tree may take from a few months to several years. Declining trees commonly set large crops of small fruit.

Roots: Black and decayed, few in number. Under healthy trees, white feeder roots are plentiful.

Under severe waterlogging, rapid decline of trees may occur. The leaves wilt and die leaving a canopy of brown dead leaves.

Source of infection and spread

The fungus causing root rot is a water mould which thrives in wet soil. Water is required for spore production, spores swimming through the soil and causing root infection.

The fungus is commonly introduced in diseased nursery trees. When the fungus is present, the environment plays a very important role in the development of symptoms. Trees may decline rapidly in sandy soils low in nutrients and organic matter but show no symptoms on deep volcanic soils. Symptoms do not appear unless there is an upset in the balance between the water requirements of the leaves and the capacity of roots to absorb water. To control root rot it is important to achieve this balance from the outset and maintain it throughout the life of the plantation.

Importance

Major disease causing substantial losses in all areas.

Control

1. Ensure internal and surface soil drainage is adequate.
2. Plant disease-free nursery trees which have been propagated in sterilised potting mixture from pathogen-free seed and irrigated with Phytopthora-free water. Such trees are available from nurseries accredited under the Australian Avocado Grower's Federation Avocado Nursery Voluntary Accreditation Scheme (ANVAS).
3. Plant trees grafted onto clonal Phytophthora-tolerant rootstocks.
4. Irrigate carefully. Irrigation systems should be capable of wetting the total root zone. Tensiometers which measure soil moisture should be used to guard against both under-and over-irrigation.
5. Increase the organic matter content in the soil. Grow cover crops between young trees. Mulch under trees with materials such as wheat straw, barley straw or sorghum stubble. Do not mulch against the trunk.
6. Use fertiliser carefully. Ideally, tree nutrition should be closely monitored by leaf analysis. Phosphorus, calcium and boron are particularly important for root growth and satisfactory levels of these elements must be maintained.

Fungicide application — field treatment of established plantations

Remove weeds from around and under the trees. Apply metalaxyl 50G granules. Apply when rain is imminent or use irrigation to ensure that the fungicide is carried into the root zone of the tree. Before the fungicide is applied to trees showing slight decline, remove all fruit and lightly prune the branches. Cut back trees with severe decline to a framework of main branches and paint the exposed limbs with a white plastic paint to prevent sunburn.

Fungicide application — replanting infested sites

Apply metalaxyl 50G granules at planting and again 8 to 12 weeks later. Irrigate to wash the fungicide into the root zone.

Fungicide foliar spray treatment of established plantations — preventative treatment

Spray healthy foliage with fosetyl systemic fungicide at 4 g/L of water. Apply 10 to 15 L/mature tree during the spring flush and again at intervals of six weeks until autumn. Apply lower volume for younger trees. Fosetyl must not be mixed with the copper fungicides used for the control of anthracnose. Phytotoxicity may occur if copper fungicide residues are present on leaves or fruit at time of application. If applied shortly after a dimethoate fruit fly spray, considerable leaf fall and fruit burn may occur.

Trunk injection

Inject 15 mL of potassium phosphite per metre of canopy diameter using pressurised syringes. The diameter of the tree is taken from the extremeties of green tissue. In severely affected trees this may be from inside the skeletal framework. Limit each syringe to a maximum of 25 mL of solution. Syringes should be placed in the main trunk of the tree and spaced evenly around the circumference of the trunk. If syringes cannot be placed in the main trunk of the tree, injection sites should be as close to the junction of the main limbs as possible.

Inject trees in early spring and repeat the treatment when the spring growth flush has matured.

Ideally, inject trees between 6 am and 11 am when the transpiration rate is highest and hence uptake is faster.

Note:

Do not inject trees in cold weather.

Do not inject sites where the trunk is damaged, for example sunburnt.

Do not prune back trees immediately before or after treatment.

Do not inject immediately above or below previous injection sites.

Thoroughly clean drill-bits and syringes between tree injections with sodium hypochlorite(1.5%) to prevent the spread of the sunblotch viroid.

Sooty blotch

Undetermined

Symptoms

Black superficial blemish on branches, stems, leaf veins and fruit.

Source of infection and spread

Spores produced during wet weather from superficial fungal growth on the cuticle of the plant.

Importance

Adequately controlled by field spray programme for anthracnose. If not controlled the disease may reduce photosynthesis of leaves and reduce the market value of affected fruit.

Control

Maintain copper fungicide programme used for anthracnose control.

Verticillium wilt

Verticillium dahliae

Symptoms

Leaves wilt and turn brown but remain attached to the tree for many weeks. In older trees, often only one branch is affected but young trees are occasionally killed. If the bark of affected branches is peeled away, a dark brown discolouration of the underlying wood is seen.

Source of infection and spread

The fungus is a common soil inhabitant and may survive for many years. It invades the tree through the roots and grows into the water-conducting tissues of the trunk and branches. Cool weather favors disease development. Many crop plants and weeds are susceptible and avocados planted in land previously used for tomatoes are commonly affected. Older trees generally recover but symptoms may reappear in subsequent years.

Importance

Minor. More likely to be a problem in land that has previously grown tomatoes or other highly susceptible crops.

Control

Avoid land where highly susceptible crops have been grown. Prune out affected branches after shooting of buds and burn prunings. When selecting grafting wood, avoid trees which have displayed symptoms of the disease.

Viroids

Sunblotch

Avocado sunblotch viroid

Symptoms

Leaves: a marked pink or white mottle or variegation. This symptom is not always expressed.

Twigs and branches: streaking and spotting of the bark. This is generally yellow but may be orange, white or almost colourless.

Fruit: yellow or red, depressed streaks generally extending from the stem- end. Affected fruit may be small and misshapen.

Source of infection and spread

The viroid is transmitted in seed and scion wood by natural root grafting between trees and possibly by pruning and cutting implements. Trees may carry sunblotch viroid without showing symptoms. Infected seeds used for producing rootstocks may cause symptoms to be expressed in the scions, even though the rootstocks themselves may show none.

Importance

Minor. Readily avoided by use of viroid-free budwood and seed.

Control

Destroy affected trees. Use only budwood and seed for rootstocks from trees known to be free from viroid. There is no absolute test for freedom from sunblotch infection but a biochemical test known as cDNA-probe is used to eliminate trees that may be symptomless carriers of the disease. Trees raised in accordance with the Australian Avocado Grower's Federation Virus-tested Tree Registration Programme (VTTR) are recommended for planting. Such trees bear a label indicating the trees have been raised in accordance with the programme. Plant these trees at least 15 m from untested avocado trees to avoid natural root graft transmission of sunblotch. Reserve a set of pruning and harvesting implements for use exclusively on virus-tested trees or thoroughly clean cutting tools used on any tree with sodium hypochlorite (1.5%).

POST-HARVEST

Bacteria and mycoplasmas

Bacterial soft rot

Erwinia sp.

Symptoms

Externally, the fruit has a darkened metallic sheen. Internally, the flesh is grey and softened, sometimes liquid and smelling rancid.

Source of infection and spread

Infection occurs in the field but more commonly after harvest. Favoured by cool wet weather as the fruit nears maturity and by harvesting on rainy days.

Importance

Sporadic but has caused significant post-harvest losses when fruit is harvested during wet weather. Fuerte and Sharwil appear more susceptible than Hass.

Control

Harvest fruit carefully to prevent surface injuries. Clip the fruit from the tree and leave the fruit stalks intact during packing. Avoid harvesting fruit during wet weather or washing them in dip tanks before packing. Chlorinate the washing water if washing is unavoidable.

Fungi

Anthracnose

The major post-harvest disease (see p. 7).

Stem-end rot

Dothiorella spp.

Symptoms

A dark brown to black rot of the fruit commencing at the stem-end as a firm, dark brown ring and proceeding as internal streaking to the other end.

Source of infection and spread

The fungus survives on dead leaves and branches where it produces large numbers of spores. During wet weather, these are deposited on fruit where infection occurs. The rot does not appear until the fruit have matured and started to soften.

Importance

Does not occur as often as anthracnose but has caused significant losses. Control using a combination of regular field sprays, post-harvest fungicide treatment and controlled ripening.

Control

1. Spray with recommended fungicides.
2. Prune out dead twigs and branches before flowering. Also remove dead leaves entangled in the tree canopy.
3. Handle fruit carefully to avoid damaging the skin.
4. Prochloraz (450 g/L) as an overhead post-harvest spray of 55 mL/100 L for 30 seconds will provide some control of stem-end rot. Controlled ripening at 16 to 18°C in combination with prochloraz treatment will greatly enhance stem-end rot control. See temperature recommendation for control of anthracnose.
5. Fruit with a short (1 to 3 days) green life (the time between harvest and when ripening commences) should be cooled promptly after harvest to 7°C. Lower temperatures after harvest could cause chilling injury.

Bananas

PLANTING MATERIAL

Fungi

Ceratocystis rot

Ceratocystis paradoxa

Symptoms

Extensive brown or dark water-soaked areas in the corm causing breakdown of tissue. A dirty white mould is generally visible in cavities in the affected tissues.

General

The fungus is a soil inhabitant which enters the corm mainly through wounds.

Importance

Minor.

Control

Do not plant affected planting material.

FIELD DISEASES

Fungi

Black pit

Pyricularia grisea

Symptoms

On fruit, small red spots enlarging to shallow black pits 3 to 6 mm wide. Pitting is most abundant on the upper hands of a bunch and on fingers on the exposed side of a bunch. The centres of mature pits may split. May be confused with damage from fruit-spotting bug.

Importance

Minor. Controlled by leaf spot and speckle spray programme.

Control

The spray schedule recommended for leaf spot will control this disease.

Black Sigatoka

Mycosphaerella fijiensis var. *difformis*

Symptoms

Both black Sigatoka and Sigatoka (caused by *Mycosphaerella musicola*) cause narrow streaks parallel to the leaf veins. Early lesions of black Sigatoka are reddish-brown and found on the underside of leaves; Sigatoka lesions are yellowish-green and more evident on the upper sides of leaves.

Lesions of both fungi develop into dark brown to black lesions with water-soaked borders surrounded by a narrow yellow band. The centres dry out, become depressed and coloured light grey, buff or beige. Many spots will cause large dead areas or leaf collapse.

General

The black Sigatoka fungus is more virulent, has a shorter life cycle and is harder to control than the Sigatoka fungus. Spread and infection are favoured by wet windy weather. The spores (conidia and ascospores) can be spread over large areas by the transport of leaves and sword suckers.

Importance

Currently this disease is restricted to non-commercial areas in the Cape York and Torres Strait regions.

Control

More frequent spraying with fungicides used to control Sigatoka disease may be required if the disease reaches commercial growing areas.

Cigar end

Verticillium theobromae

Symptoms

A firm dark decay at the end of fruit surrounding the dead floral parts and extending back along the fruit up to 20 mm. The affected area shrinks and becomes more or less rounded in contrast to the angular shape of immature fruit. The area later becomes covered with an ashy grey coating of spores resembling a burnt cigar tip. The fruit pulp undergoes a dry rot. Generally only a few fruit on the hand are affected.

Source of infection and spread

The fungus colonises banana flowers and leaf trash and infects fruit during wet weather.

Importance

Minor.

Control

Not generally warranted.

Cordana leaf spot

Cordana musae

Symptoms

When mature, spots are elliptical or oval-shaped, up to 100 mm long, brown with zonate patterns on the upper surface and grey underneath, and bordered by a conspicuous bright yellow band.

Source of infection and spread

Large numbers of spores produced on the underside of affected leaves are spread during wet windy weather.

The fungus often invades leaf tissues weakened by leaf spot (Sigatoka disease).

Importance

Minor. Controlled by leaf spot and speckle spray programme.

Control

Spray with recommended fungicides.

Fungicide sprays

See Leaf spot section (p. 12).

Fruit speckle

Deightoniella torulosa

Symptoms

On fruit, sunken dark brown or black spots, 2 to 4 mm wide, are surrounded by a prominent dark green halo. The disease may be present on all bunches at all stages of maturity and is more prevalent on the outside and towards the blossom end of the fruit.

Source of infection and spread

Spores of the fungus produced on dead banana leaves during wet weather are discharged into the air and land on fruit where speckling develops within a few days. The fungus does not produce spores on the fruit.

Importance

Minor. Controlled by leaf spot and speckle spray programme.

Control

Specific control measures are not warranted if leaf spot and speckle control recommendations are implemented.

Leaf speckle

Mycosphaerella musae

Symptoms

A general yellowing of leaves with dark brown or black speckled patches on the undersides. Speckling generally does not appear above the fifth or sixth youngest leaf on a plant before flowering. Fruiting bodies of the fungus are visible on the patches. When the disease is severe, loss of leaves results in smaller bunches. Fruit ripen early and unevenly, are poorly filled and more susceptible to sunburn.

Source of infection and spread

Infection occurs from spores (ascospores) produced in small black fruiting bodies (perithecia) on old leaves. These are carried by wind to the developing leaves where they germinate. The fungus then grows on the leaf surface for five to six weeks before penetrating. For this reason, infection is generally confined to the undersides of leaves which are shaded and remain moist longer.

Importance

Can be serious, particularly in north Queensland, but can be controlled by a regular spray programme.

Control

1. Spray with recommended fungicides.
2. Destroy badly speckled leaves and trash.

Fungicide sprays

See Leaf spot section (p. 12).

Leaf spot-Sigatoka

Mycosphaerella musicola

Symptoms

The heart leaves are usually free from visible spotting until flowering and leaf production has ceased.

Before flowering, the third or fourth leaf from the centre shows light yellow or brown green streaks 1 to 2 mm long, parallel to the veins. The fifth and sixth leaves show muddy brown to black elliptical spots up to 100 mm long. On older leaves, spots are light grey, with a dark brown or black border often surrounded by a narrow yellow band. Small grey specks, which are tufts of fungal spores appear on the spots during wet weather. If spots are numerous, they coalesce to form large dead areas.

When the disease is severe, bunches are small. Fruit ripen early and unevenly, are poorly filled and prone to sunburn because of leaf loss.

Source of infection and spread

Large numbers of spores (conidia) are produced on affected leaves and spread during wet windy weather. Leaves are susceptible from the time they appear in the throat of the plant but do not show symptoms for some weeks. In south Queensland, infection is most likely to occur from December to March; in north Queensland it occurs throughout the year.

Another type of spore (ascospore), produced on old leaves, may also cause infection resulting in tip-spotting of young fully expanded leaves. This contrasts with the conidial infections which are scattered over the leaf.

Importance

Can be serious, particularly in north Queensland, but can be controlled by a regular spray programme.

Control

1. Spray with recommended fungicides.
2. Destroy badly spotted leaves and trash.

2. Destroy badly spotted leaves and trash.

Fungicide sprays

Besides controlling leaf spot and speckle this schedule also controls cordana leaf spot, fruit speckle and black pit.

Spray with protectant fungicides: mancozeb, or maneb plus zineb. Combine with white oil or water miscible oil.

Chlorothalonil is also recommended for control of leaf spot and speckle. When using chlorothalonil **do not** add spraying oils, wetting agents or combine with emulsifiable concentrates.

Apply in adequate water to give good coverage. If using large tractor-drawn air-blast machines apply 200 to 600 L/ha of the spray. If using back-mounted misting machines apply 50 to 100 L/ha.

During periods of high disease pressure, or following extended periods of wet weather, several applications of the eradicant fungicide propiconazole can be substituted for the protectant fungicides. Propiconazole is recommended for eradicant sprays only. This minimises the risk of resistant fungal strains developing.

Note:

1. The recommended rates of fungicide and oil per hectare **must** be maintained - only the volume of water is varied.
2. Special misting oils without emulsifiers may be used in misting machines instead of miscible oils. These 'straight' oils are normally used when fungicides are not used; that is when speckle is not present. For leaf spot and speckle control, it is necessary to use a fungicide and a miscible oil.

Marasmiellus rot

Marasmiellus inoderma

Symptoms

Withering, death and eventual decay of leaf sheaths with a conspicuous white fungal growth on affected areas. New leaves are slow to emerge and roots may be killed. Pale yellow to brown mushrooms (caps have widely separated gills) with white stalks are produced at soil level and on the surface of the pseudostem in wet weather.

Source of infection and spread

The fungus is a common organism on decaying banana trash but may also be found on weakened corms. It attacks plants which have been debilitated by adverse growing conditions, particularly drought. The disease is generally confined to poor soils.

Importance

Minor.

Control

Improve growing conditions.

Panama disease

Fusarium oxysporum f.sp. *cubense*

Symptoms

First shows as a marginal yellowing of the older leaves which later turn brown and dry out. They eventually collapse at some point along the leaf stalk or at the junction of the leaf stalk and the pseudostem, leaving a skirt of dead leaves draped around the plant. Death of the parent pseudostem generally follows but suckers do not necessarily die.

The most characteristic symptom may be seen by cutting through the pseudostem of an affected plant near ground level where a dark brown to black discolouration of the water-conducting tissues is evident. Corms when cut show brown or black lines running through the tissues.

Source of infection and spread

The disease is most commonly introduced in planting material. Once established, the fungus persists in the soil for many years. It enters the plant through the roots and grows into the water-conducting tissues of the corm and pseudostem. Spread from a diseased parent plant into its suckers also occurs.

Importance

Major disease in Lady Finger cultivars. Outbreaks of Panama disease are occuring in Cavendish cultivars in southern Queensland where a new race of *Fusarium oxysporum* f. sp. *cubense* has developed.

Control A (to avoid infection)

1. Use planting material from disease-free plantations or from tissue culture.
2. Do not plant susceptible cultivars on land where the disease has occurred.
3. Avoid a site where surface runoff from a diseased plantation can contaminate the soil.
4. Avoid poorly drained soils.

Control B (if infection occurs)

Act quickly once the presence of the fungus is confirmed, but do not dig out infected plants as this is likely to spread the fungus.

1. If only one plant is affected:

(a) Chop up all plant material from the affected stool including the gouged out corm tissue and place into large 60 L plastic bags. Add 1 kg of urea to each bag to hasten breakdowwn of fungal and plant tissue.

(b) Treat the affected site with dazomet granular soil fumigant at 100 g/m^2; water in with at least 10 L/m^2 of 10% urea solution and cover the site with plastic sheeting. Place the garbage bags containing the infected banana material on the plastic sheet to hold it in place, leaving it undisturbed for as long as possible.

(c) Inject a surrounding ring of healthy plants (two stools deep) with herbicide.

(d) Isolate the area with a permanent fence.

2. If more than one plant is affected, kill each infected stool by injecting herbicide. Inject surrounding plants and isolate the area.

Management

1. A plant with fusarium wilt may sometimes produce a marketable bunch but the fungus may be present in the bunch stalk. If this is dumped in the plantation it may be responsible for a new infection.
2. Do not move infested soil on implements, tyres or footwear.
(a) Disinfect footwear - use a copper fungicide/ methylated spirits mixture.
(b) Disinfect implements, including the legs of ladders used in the plantation, by washing free from soil and sterilising with methylated spirits, formalin or sodium hypochlorite.

Sooty blotch
Chaetothyrina musarum

Symptoms

A light brown superficial blemish of the skin of the fruit.

General

It is most likely to occur in cool wet weather.

Importance

Minor.

Control

Not warranted.

Nematodes

Burrowing nematode
Radopholus similis

Symptoms

Red, purple and black areas in the roots and corm. Affected stools generally decline prematurely, produce smaller bunches and are prone to fall in wet windy weather because of root breakdown.

Source of infection and spread

The burrowing nematode is an introduced pest not found in virgin land unless it has received run off from infested plantations. The nematode may survive on weed hosts after infested stools are destroyed. It is generally spread from one plantation to another in planting material. Within a plantation, it is carried in run-off and in soil adhering to farm workers, animals and equipment.

Importance

Major problem in all areas.

Control

1. Carefully select planting material and treat in hot water. Planting material from a QDPI-approved scheme should be used for Lady Fingers and Tall varieties, or use plants from tissue culture.
2. Spell land for two years before replanting.
3. Treat soil with recommended nematicides before and after planting.

Control before planting:
Work soil deeply to a fine tilth and treat with a registered soil fumigant in accordance with label directions. Cost may be reduced by treating a 2 m-band where the row is to be planted rather than the entire area. Cultivate in the direction of the rows to reduce the rate of reinfestation of the treated band.
Treatment of planting material:
Pare roots and discoloured tissues from the corms and immerse in hot water at 53 to 55°C for 20 minutes. After treatment, spread in a single layer in the shade to cool. Hot water tanks specially constructed for treating banana planting material are available.
Control after planting:
Apply fenamiphos or ethoprophos as per label.
Note:
1. Virgin land is generally free from burrowing nematode.
2. Old banana land should be spelled for at least 12 months before replanting and preferably two years. Volunteer banana plants must be eradicated during this time.
3. Do not plant cover crops such as maize, cowpea, sorghum-Sudan grass hybrids or stylo as these are alternative hosts of the burrowing nematode. Siratro and green panic are suitable cover crops.

Root-knot nematodes
Meloidogyne incognita and *Meliodogyne javanica.*

Symptoms

Galls or swellings on roots.

Importance

Considered to be of minor importance but may occasionally warrant control. Effects may be masked through the chemical control of burrowing nematode

Control

Refer to the control of burrowing nematode.

Root-lesion nematode
Pratylenchus coffeae

Symptoms

Blackening of the root epidermis, a reduction in the number of lateral roots and small reddish lesions in the superficial cortex. The lesions do no extend into the rhizome or butt of the stool. Affected stools are unthrifty and develop into smaller plants with small bunches. They are not a cause of toppling plants.

Source of infection and spread

It is generally spread from one plantation to another in planting material. Between and within plantations it can be carried in run-off water and in soil adhering to farm workers, animals and equipment.

The nematode can survive on weed hosts after infested stools are destroyed.

Importance

Minor, due to its limited distribution. Important in infested plantations.

Control

See burrowing nematode (p. 14).

Spiral nematode

Helicotylenchus multicinctus

Symptoms

Blackening of the root epidermis, a reduction in the number of lateral roots and small reddish lesions in the superficial cortex. In severe infections the discolouration extends deep into the cortex. The lesions do not extend into the rhizome or butt of the stool. Affected stools are unthrifty and develop into smaller plants with small bunches. They are not a cause of toppling plants.

Source of infection and spread

As for burrowing nematode (p. 14).

Importance

Minor importance due its limited distribution. Important in infested plantations.

Control

Refer to control of burrowing nematode.

Viruses

Bunchy top

Banana bunchy top virus

Symptoms

The most obvious symptom is the development of dark green flecks along veins of the leaves producing a 'dot-dash' pattern. This is best seen at the base of the leaf adjacent to the midrib. The dark green flecks in the veins can often be seen 'hooking' into the midrib from the leaf blade.

Affected leaves are more upright and become pale yellow around the margin. The margins are also more wavy than normal. Growth is reduced and the emerging leaves become choked in the throat of the plant producing a 'bunchy top' effect. Affected plants are stunted and vein flecking can be seen in the leaf sheaths of the pseudostem as well as in the leaves. Plants affected at an early stage of growth rarely produce bunches. Those affected later may produce bunches which project upwards from the throat of the plant rather than hang. Flecking can frequently be seen in the flower bell.

Source of infection and spread

The virus is spread from diseased to healthy plants by the banana aphid (*Pentalonia nigronervosa*). Aphids feeding on diseased leaves retain the virus for long periods and, if wind-borne, may spread the disease over long distances.

The virus may also be spread in diseased planting material.

Importance

Major disease in south Queensland. The incidence is usually low as a result of the regular inspection and eradication programme.

Control

1. The infected plant must be sprayed immediately it is found, with either kerosene or mineral oil, to kill all the aphids on it.
2. The infected plant, and all its attached suckers, must then be destroyed within 48 hours by one of the following two methods:

(a) The complete stool may be dug from the soil, and the underground parts chopped into pieces not exceeding 50 mm in diameter. This will avoid any possibility of regrowth occurring

or

(b) The plant and each of its suckers may be injected with a 1% solution of 2,4-D at a point on the pseudostem 300 mm above the base of the plant. Any subsequent sucker regrowth must be destroyed by spraying with a 0.2% solution of 2,4-D.

Queensland growers should refer to the Department of Primary Industries about eradication requirements within the designated quarantine areas.

3. Use planting material only from a plantation which has been free from bunchy top for at least the preceding two years.

Note: Any grower who finds bunchy top on his property is required to notify the local banana inspector promptly.

Infectious chlorosis

Cucumber mosaic virus

Symptoms

A yellow streaking or mosaic pattern on the leaves, often with a curling of the margins. Rotting of heart leaves and the centre of the pseudostem may occur in cool weather.

Source of infection and spread

The virus has a very wide host range, including many weeds, from which it is spread to bananas by aphids. Spread between banana plants is very limited. The disease may also be introduced in planting material.

Importance

Minor.

Control

None warranted.

POST-HARVEST

Fungi

Anthracnose

Colletotrichum musae

Symptoms

Anthracnose: on ripe fruit, circular brown sunken spots which become covered with a pink to rusty-red mass of spores. Spots may coalesce to cover large areas of the fruit.

Black end: blackening and rotting of the cut ends of single fruit. As the fruit ripens, the rot advances rapidly down the stalk into the flesh.

Crown rot: blackening and rotting of the cut ends of the hands. As the fruit ripen, the rot advances rapidly down the fruit stalks causing them to fall from the hand.

Source of infection and spread

The disease is a post-harvest problem and rarely affects fruit in the plantation. Spores of the fungus are produced in large numbers on dead banana leaves and are spread by wind-driven rain to green fruit. After infection, the fungus remains dormant in the tissues until ripening. Black end and crown rot develop after individual fruit and hands have been cut from the bunch.

Importance

The most serious post-harvest disease but can be controlled by fungicide treatment.

Control

1. Dip or spray fruit with a recommended fungicide.
2. Handle fruit carefully after harvest.
3. Keep packing sheds free from rotting fruit and banana trash – remove daily.
4. Steam clean or wash shed and equipment at least once a week with high pressure water and detergent followed by an aqueous solution of either 2 mL of 10% sodium hypochlorite or 2 mL quarternary ammonium compound/L or 20 to 50 mL of 40% formalin/L. **Warning:** If using formalin, follow recommended saftey procedures.
5. Keep fruit cool after harvest.
6. Detrash plants regularly.

Post-harvest dip

After dehanding and washing, dip in benomyl or carbendazim plus a wetting agent as recommended by the manufacturer for one minute **or** thiabendazole plus a wetting agent as recommended by the manufacturer for two to four minutes or apply prochloraz as a non-recirculated spray for two minutes.

Note:

1. Wettable powder fungicides should be made into a slurry in hot water before being added to the dipping tank. Do not add the powder directly.
2. Thoroughly stir the dip before and during use.
3. Top up to the original volume regularly using a suspension equivalent in strength to that of the original dip.
4. Use one lot of dip for each 2500 kg of fruit and then discard. If excessive dirt, sap and trash accumulate in the dip, replace sooner. The dip will remain effective for at least one week and possibly longer.
5. A suitable spraying system for the application of prochloraz consists of two TX2 hollow cone nozzles mounted in series above the fruit on a packing line. At an operating pressure of 320 KPa (46 PSI) the total delivery rate is 16 L/hr. Two TX2 nozzles are favoured over 4TX1 nozzles which deliver the same volume but clog easily.

Ceratocystis fruit rot

Ceratocystis paradoxa (conidial state: *Thielaviopsis paradoxa*).

Symptoms

An uneven black rot of hands of fruit extending from the crown. A white to grey fungal growth forms on affected areas and a sweet smell is present. Once finger stalks are affected, fingers readily drop from the hand.

Source of infection and spread

The disease is a post-harvest problem, infection occurring through the cuts where the hands are removed from the bunch. The main sources of infection are old bunch stalks and discarded fruit in and around the packing shed.

Importance

Generally minor but some serious outbreaks have occurred in wet seasons.

Control

Remove banana refuse regularly from the packing shed.

Squirter

Nigrospora sphaerica

Symptoms

The pulp of the fruit decomposes to a dark, semi-fluid state which is easily expelled from the end or side of the fruit when squeezed. There may be no external indication of the rot but severely affected fruit may have a bluish- black bruised appearance over the badly affected parts. Occasionally, the finger stalk may show a soft dark rot.

Source of infection and spread

Squirter is a post-harvest disease and is not seen in the field. The fungus enters through the broken finger stalk.

Importance

Generally controlled by post-harvest fungicide treatment. Squirter may be a problem in fully ripe fruit sent to interstate markets during winter. The disease may also occur in chilled fruit that has undergone an extended ripening period.

Control

Use the recommended post-harvest treatment of benomyl (40 g/100 L).

Citrus

FIELD DISEASES

Fungi

Alternaria leaf spot

Alternaria alternata - lime and rough lemon strain

Symptoms

Light to dark brown spots develop on the leaves surrounded by an irregular yellow halo. Spots vary considerably in size and may involve large areas of the leaves. Affected leaves fall readily and shoot dieback may follow.

Importance

Minor.

Control

The disease is common in Queensland on rough lemon seedlings but control is not warranted since rough lemon is used as a rootstock only.

Black spot

Guignardia citricarpa

Symptoms

Leaves: small slightly sunken spots up to 3 mm wide may occasionally occur.

Fruit: four distinct types of symptoms occur:

(a) speckled blotch - localised circular areas, dark brown to black speckling of the skin. Develops on green fruit when late summer and early autumn weather is abnormally hot and dry.

(b) freckle spot - deep orange to brick red, slightly depressed spots up to 1 mm wide. These enlarge to produce the hard spot symptom described below.

(c) hard spot - circular, depressed spots up to three mm wide, with grey to white centres and black margins. A ring of dark green tissue may surround the spot when the fruit is mature. The spots contain small black fruiting bodies of the fungus.

(d) virulent spot - occurs when spots expand rapidly and coalesce, sometimes covering up to half the surface. Large numbers of small, fruiting bodies which develop give these areas a dark speckled appearance.

Source of infection and spread

Most infection results from primary spores (ascospores) ejected from fruiting bodies (perithecia) which develop on fallen citrus leaves on the orchard floor. These are carried by wind and infect young leaves and fruit. The fungus then ceases to grow and remains dormant in the tissues. Leaf symptoms rarely show until leaf-fall when the fruiting bodies begin to develop. Fruit generally first show symptoms at ripening, particularly during warm to hot weather. It is seen first on the exposed sides of fruit on the warmer northern side of the tree. Fruit is susceptible from just after petal-fall for a period which depends on the cultivar and the age and vigour of the tree. Secondary spores (conidia), produced on the fruit spots and spread by water to young fruit, are of little importance.

Importance

Serious in coastal orchards in wet seasons, particularly in late-maturing cultivars such as Valencias and Murcotts. The standard domestic spray schedule provides satisfactory control in inland areas, however, more stringent control measures are required for Valencia oranges for export to Japan.

Control

Spray with the recommended fungicides.

Spray programme for domestic market

Stage: late September to early October (or 1/2 to 3/4 petal fall)

Chemicals: copper oxychloride or copper hydroxide. Add white oil, miscible summer oil, or narrow range oil.

1. Premix all ingredients separately with a little water.
2. Add each to the spray vat with the agitator working.
3. Add the oil last.
4. Two or more half-strength sprays two weeks apart give better protection when blossoming is protracted as in lemons or wet weather occurs at flowering.
5. This spray helps to prevent stem-end rot.

Stage: mid November (or six weeks after petal fall spray).

Chemicals: mancozeb, copper oxychloride or copper hydroxide. Add white oil, miscible summer oil or narrow range oil.

1. Wet windy weather promotes wind-rub damage to young fruit, and darkening of fruit blemishes is accentuated by copper.

Stage: late December (or 12 weeks after petal-fall spray).

Chemicals: as for mid November sprays.

1. Copper sprays (not mancozeb) help prevent Phytophthora brown rot, collar rot and pink disease and may reduce the need to spray specifically for these diseases.

Expanded spray programme for valencia oranges for export to Japan

The 'nil tolerance' to black spot imposed by Japanese citrus importers demands more stringent control measures to be applied by citrus growers.

Stage:

1/2 to 3/4 petal fall; copper, six weeks later – mancozeb or copper

12 weeks post-petal fall; mancozeb or copper

16 weeks post-petal fall; mancozeb

20 weeks post-petal fall; mancozeb + benomyl

(a) If flowering is extended, apply two copper sprays, at the full rate, no more than 14 days apart during flowering.
(b) No more than two copper sprays to be used after petal fall sprays in any one season.
(c) Benomyl used pre-harvest on Valencias is optional rather than obligatory, especially in young orchards, where black spot has not been recorded.
(d) Complete coverage of fruit at each spray application is required.

The appropriate rates of chemical product to be used:

	Mid-low volume (g/100 L)	High volume* (kg/ha)
copper oxychloride + +	400	10
mancozeb +	200	5-10
benomyl +	50	1.25-2.5

* HV spray rates to be used where a volume of 20 L/tree (which is equivalent to 4000 L/ha) is exceeded.

\+ Rate refers to product, not active ingredient, per 100 L or per ha.

\+ + These rates refer to copper oxychloride 500 g/kg WP formulations. Other formulations of copper oxychloride or copper hydroxide may be used at the label rates for black spot.

Brown rot

Phytophthora citrophthora

Symptoms

Leaves: symptoms include blackening and death of young tip growth. Dark, water-soaked areas occur on the leaves, particularly at the tips and along the margins. Affected leaves often fall while still green.

Fruit: symptoms are first evident as a dull brown discolouration of the rind, on one side of the fruit. The affected area remains more or less circular as it enlarges and may cover the entire fruit. The rot remains firm and becomes darker in colour and slightly sunken. Affected fruit have a characteristic pungent odour and fall readily. In wet weather, a white fungal growth appears on the surface of the rotted area.

Source of infection and spread

The fungus causing brown rot is a soil-inhabiting water mould and requires water for spore production and fruit infection. It is most active at temperatures near 25°C.

Importance

Brown rot is troublesome in coastal districts during wet seasons.

Control

1. Spray with the recommended fungicides before the onset of late summer-autumn rains. If prolonged wet weather occurs, a second spray will be required as soon as weather permits. Spray the fruit and foliage below a height of 1.5 m, the ground under the tree and the trunk.
2. Control weeds under trees and remove branches less than 300 mm above the ground to permit better air circulation and reduce spore splash to leaves and fruit.

Chemical: copper oxychloride.

1. Copper sprays applied for black spot control may provide adequate protection against brown rot during the summer months. Heavy rain during January and February necessitates brown rot control before autumn.

Brown spot - Mandarins

Alternaria alternata

Symptoms

Leaves: infected areas become black water-soaked spots, often restricted to the leaf margins. Leaves often curl around the affected areas and may fall.

Twigs: blighting and blackening of young growth.

Fruit: initially, small brown sunken spots develop enlarging up to 5 mm wide; the centre may become slightly raised and protrude. However, some small spots do not enlarge and may dry out leaving a corky cracked scar. Affected fruit colour prematurely and may fall.

Source of infection and spread

The fungus survives on old infected twigs. Large numbers of spores are produced and are spread by wind and water to leaves and fruit.

Importance

A serious disease of Murcott and Emperor mandarins.

Control

Spray with recommended fungicides. Spray new leaves and fruit to protect them from infection. Start when the first leaf flush is 3/4 expanded. Apply a protectant spray to each new growth flush and developing fruit until April.

Chemicals: copper oxychloride plus zinc sulphate heptahydrate plus caustic soda plus white oil. Iprodione is registered for use on non-fruiting trees.

1. Ensure that new leaves on young growth are fully protected; fully expanded leaves are resistant to attack.
2. Iprodione does not control melanose, black spot, Phytophthora brown rot, smoky blotch or scab.
3. High-volume spraying for thorough coverage of new growth flushes and fruit is recommended.

Crotch rot - Mandarins

Diaporthe citri

Symptoms

A rot in the crotches of the trunk and branches. The bark cracks and, in the advanced stage, may loosen in vertical strips. Foliage becomes yellow, and premature defoliation occurs, leaving bare dead twigs.

Source of infection and spread

Spores of the fungus produced on dead twigs and branches are washed into the crotch by rain and irrigation water.

Importance

Serious in Hickson mandarins in coastal areas.

Control

Do not grow Hickson mandarin in humid coastal areas.

Melanose

Diaporthe citri

Symptoms

Leaves: reddish brown to black, raised pustules develop. Affected leaves are rough to touch and may twist and fall prematurely.

Fruit: minute, reddish brown to black spots, generally appear as 'flyspecks' but are often arranged in patterns such as circles and 'tear stains'. Affected areas are rough to touch. Spots may be so dense as to form a crust which cracks as the fruit enlarges to produce the symptom known as 'mudcake' melanose.

Source of infection and spread

The fungus produces large numbers of spores on dead twigs and branches. These are spread to adjacent leaves and fruit by splashing and dripping water, hence the circular and 'tear stain' patterns produced on fruit. Leaves and fruit are susceptible to infection only when very young, most fruit infection occurring within three weeks of petal fall.

Importance

Melanose is of major importance in coastal areas but minor elsewhere.

Control

1. Spray with the recommended fungicides at 1/2 to 3/4 petal fall. If flowering is prolonged, a second spray is required two to three weeks later.
2. Prune out dead twigs and branches prior to flowering.

Stage: Late September to early October (or when 1/2 - 3/4 petal fall).

Chemicals: copper oxychloride or copper hydroxide. Add white oil, miscible summer oil, or narrow range oil.

1. Premix all ingredients separately with a little water.
2. Add each to the spray vat with the agitator working.
3. Add the oil last.
4. Two or more half-strength sprays two weeks apart give better protection when blossoming is protracted, as with lemons or when showery weather occurs during flowering.
5. This spray also helps to prevent stem-end rot.

Phytophthora collar rot

Phytophthora nicotianae var. *parasitica* and *Phytophthora citrophthora*

Symptoms

Root rot: a tan to dark brown rot of feeder roots which extends into the larger secondary and primary roots causing extensive discolouration and necrosis of the root system. Yellowing of the leaves, defoliation and twig die-back is a common consequence of root rot. One branch or side of the tree may be more seriously affected than the others depending on the extent of root rot in the subtending roots.

Collar rot: the main symptoms are initially dark water-soaked areas of bark overlaying discoloured cambium and sapwood tissue, often with gum exudation on the trunk near ground level. On Phytophthora-susceptible rootstocks infection usually progresses from the major roots up the trunk, although direct infection of the trunk can occur through wounds. Dead bark usually supports pinkish-white fusarium mould growth and later desiccates, cracks and breaks away from the wood in vertical strips. On cutting through the bark a distinct brown discolouration is evident in the underlying wood, and this may extend beyond the area of the external bark symptoms. The margin between diseased and healthy wood is distinct. Leaves of trees with collar rot are often yellow.

Source of infection and spread

The fungi causing these diseases are soil-inhabiting water moulds requiring water for spore production and root infection.

The optimum temperature for *Phytophthora citrophthora* is 4 to 5°C lower than that for *Phytophthora nicotianae* var. *parasitica.*

Importance

Particularly serious in wet coastal orchards and in inland areas with irrigated shallow marginal soils with poor internal drainage. Rough lemon, sweet orange and mandarin rootstocks are very susceptible to root and collar rot.

Control

1. Plant disease-free nursery trees.
2. Use deep well-drained soils.
3. Avoid overwatering trees.
4. Do not plant nursery trees in the field deeper than they were in the nursery.
5. Avoid damaging bark during cultural operations.
6. Control weeds around trees and raise skirts of trees to permit air movement around the trunk.
7. Do not place mulch against the trunk.
8. Use tolerant rootstocks such as Troyer citrange particularly in replant areas.
9. Apply a copper oxychloride suspension as a thin paste after carefully cutting away all discoloured bark and wood tissues and at least 10 mm of bark beyond discoloured area. Once callus tissue forms, paint excised area with a tree-sealing

compound, for example, Colgraft or white plastic paint.

Retreatment of trees may be required. Inspect for new sites of bark rotting and gummosis and treat promptly.

Pink disease

Corticium salmonicolor

Symptoms

Wilting and collapse of individual branches occurs as a result of branch-girdling by a pink (becoming white then grey with age) fungal growth on the bark and cambium tissue, particularly in wet weather. As the branch is girdled, extensive gum exudation may occur. The bark splits and lifts off before the branch finally dies. Unless checked, the fungus spreads through the limbs killing large sections of the tree.

Source of infection and spread

The fungus attacks many woody species besides citrus. It is most common in humid tropical areas but also occurs in moist subtropical coastal areas, particularly in areas adjacent to rainforest.

Importance

Serious in wet tropical areas.

Control

1. Prune out diseased branches as soon as they are detected. Paint cuts with the recommended fungicide and burn prunings.
2. Once the disease has been observed inspect the orchard regularly, particularly during the summer months.
3. Apply the fungicide copper oxychloride before the onset of the wet season and repeat at least once during the season. Spray the entire tree including the trunk.

Scab

Elsinoe fawcettii

Symptoms

Raised grey to light brown corky scabs are the main symptoms on leaves, twigs and fruit. Affected leaves may be distorted.

Lemon and mandarin cultivars are most susceptible. The disease is a serious nursery problem particularly on rough lemon seedlings.

Source of infection and spread

Old scabs on leaves, twigs and fruit provide the main source of infection. Spores produced on these scabs are spread by wind and rain. Only young tissues are susceptible.

Importance

Scab is a minor disease usually well-controlled by the standard spray programme for melanose and black spot.

Control

Spray with the recommended fungicides.

Refer to the spray programme recommended for melanose control.

Smoky blotch

Gloeodes pomigena

Symptoms

Light brown irregular blotches occur on the skin of fruit.

Source of infection and spread

The damage is very superficial but may cause a down-grading of fruit quality.

Importance

Minor.

Control

The spray programme recommended for black spot also controls this disease.

Sooty mould

(miscellaneous dark-coloured fungi)

Symptoms

A dark fungal growth on leaves, twigs and fruit where scale insects or aphids have fed. This mould affects the colouring of the fruit.

Source of infection and spread

The fungi causing sooty mould are not parasitic but entirely superficial and live on the honeydew secretions of scale insects and aphids.

Importance

The disease is minor if scale insects and aphids are controlled satisfactorily.

Control

Use the recommended spray programme for insect pests.

Nematodes

Citrus nematode

Tylenchulus semipenetrans

Symptoms

Affected trees appear stunted and decline slowly. Examination of the roots offers the best means of diagnosis. Because of the gelatinous material in which nematode eggs are laid, soil particles cling to infested roots even when washed under running water; whereas soil particles are easily removed from healthy roots.

Source of infection and spread

The nematode may be introduced in infested planting material and can be spread in soil carried by surface water or machinery.

Importance

Widespread. Reduces growth and yield.

Control

Citrus nematode does not occur in virgin soils and every effort should be made to keep new plantings free of the nematode.

1. Buy nematode-free trees.
2. Thoroughly clean all cultivation equipment before moving it from an old orchard to newly planted orchards on virgin soil.
3. Ensure that irrigation or run-off water from infested blocks does not contaminate non-infested blocks. Citrus nematode is particularly damaging in replant situations because young trees are exposed to high nematode populations that carry over from the previous crop.
4. Use resistant rootstocks in replant blocks or shallow soils (less than 750 mm deep). Troyer citrange, widely recommended as a replant rootstock, has reasonable resistance to nematodes.
5. Let replant soil lie fallow for as long as possible to reduce the nematode population. For example, a two-year fallow will substantially reduce nematode numbers.
6. If susceptible rootstocks must be used and replant land cannot be fallowed, fumigate the soil with methyl bromide. Contractors will fumigate large areas with methyl bromide.

Good management practices such as regular watering, mulching, adequate fertilisation and timely control of other pests and diseases reduce the stress on citrus trees and help reduce the effects of citrus nematode.

Undetermined cause

Shell bark of lemons

Symptoms

Cracking and peeling of the bark on the trunk and main limbs occurs with a variable amount of gumming. The outer bark dies and peels off in strips but the inner bark stays alive and replaces the outer layer. After healing, the condition may not recur for some years. Growth may be checked and considerable damage results when subsequent fungal infection occurs. In Queensland most Villa Franca, Lisbon and Eureka lemons are affected.

Importance

Minor but common in mature lemon trees.

Control

1. Heavy pruning appears to aid recovery.
2. Do not take budwood from affected trees.

Viruses/viroids

Psorosis

Citrus psorosis virus

Symptoms

Patches of bark on the trunk and main limbs become scaly. The bark lifts in irregular flaps which remain attached to the tree in the centre or on one side. The wood under affected bark dies.

Source of infection and spread

An insect vector is not known and the virus is not seed-transmitted. Symptoms are rarely seen in trees under 10 years of age.

Importance

Rare.

Control

Affected trees cannot be cured but their life may be prolonged by pruning out badly affected limbs, scraping away diseased bark from less affected areas and painting with the recommended wound dressing. Control is achieved by using budwood from trees not showing symptoms of psorosis. Use indexed budwood from the citrus budwood scheme.

Scaly butt - exocortis

Citrus exocortis viroid

Symptoms

Cracking and scaling of the bark occurs below the bud union with the bark, often peeling off in vertical strips. Affected trees are stunted and the trunk below the union is reduced in size. Symptoms generally become obvious only in trees more than four years old.

Source of infection and spread

The viroid is spread by using buds from infected trees. It is apparent only on trees worked on *Poncirus trifoliata* and citrange rootstocks. However, other rootstocks may carry the viroid without showing symptoms.

Importance

Scaly butt can be serious if indexed budwood is not used.

Control

When using *Poncirus trifoliata* and citrange rootstocks, take budwood from healthy mature trees already on those rootstocks. Use indexed budwood from the citrus budwood scheme.

Stem pitting - grapefruit
Citrus tristeza virus

Symptoms

Pitting and grooving of the wood surface occurs and is usually seen only when the bark is removed. In severe cases, the trunk and larger branches are dimpled and distorted. Affected trees are stunted and bushy, have small fruit, twig dieback and a yellow mottling of leaves during growth flushes. Occasionally trees decline rapidly.

Source of infection and spread

The virus is spread through budwood from affected trees and by the common black citrus aphid. All popular grapefruit cultivars, particularly Marsh, are susceptible, and the disease develops irrespective of the rootstock.

All citrus trees carry the virus but only grapefruit show obvious symptoms.

Importance

Major in grapefruit.

Control

1. Select budwood from trees carrying only mild strains of the virus.
2. Topwork affected trees to other types of citrus.

POST-HARVEST

Fungi

Blue and green mould
Penicillium italicum (blue mould) and *Penicillium digitatum* (green mould)

Symptoms

Fruit only affected. Small soft water-soaked areas which develop rapidly in warm humid conditions, eventually rotting the whole fruit. Rotted areas become covered with a dense white fungal growth from the centre of which masses of spores develop to give the mould its characteristic colour. Spores rise in clouds when affected fruit are disturbed.

Blue mould grows much more slowly than green mould and the blue spore masses keep pace with the white fungal growth. It may also produce spores inside rotted fruit.

Source of infection and spread

Blue and green moulds are mainly post-harvest diseases and are found in the orchard on damaged fruit only. The main source of the fungus is diseased fruit in and around the packing shed. Infection occurs through injuries to the skin although many of these are small and scarcely visible. Even a ruptured oil gland provides enough nutrients for spores to germinate and infect the fruit. In containers, the fungus may spread from fruit to fruit by contact, forming 'nests' of rotting fruit. Mould growth is rapid at 20 to 25°C and very slow below 5°C and above 30°C.

The moulds are worst in wet weather when the fruit is easily damaged and conditions are ideal for the spread of the fungus.

Importance

Moulds are a major problem where post-harvest treatment is not used.

Control

Hygiene and everything possible to minimise fruit damage during picking and handling should be implemented.

1. Bury mouldy fruit.
2. Pickers should have short fingernails and preferably wear gloves.
3. Use correct picking methods.
4. Cut stems close to button with properly maintained clippers.
5. Roll fruit gently into bulk bin.
6. Use clean bulk bins, free of splinters.
7. Avoid jolting during transport to sheds.
8. Place bins in shade.
9. Do not hold untreated fruit in packing sheds for longer than 24 hours.
10. Remove mouldy fruit and bury daily.
11. A poorly maintained packing line may contribute to more than half the rind injuries during harvesting and packing.
12. Repair worn or defective brushes.
13. Sorters and packers should wear gloves.

Fungicides

Management procedures depend on (a) how long fruit is held before packing (options 1 and 2), and (b) whether it is to be exported. Apply fungicide dip and spray treatment within 24 hours after harvest.

Option 1. If holding fruit more than 24 hours before a packing-line treatment is to be applied.

Fruit should be bulk-dipped immediately after harvest in guazatine or, where permitted, imazalil (not for fruit being exported to Japan). After dip treatment, fruit can be safely stored or degreened for several days before being put through packing line treatment.

Note: the fungicide used in the packing line should belong to a chemical class differing from the one used for bulk dipping. This strategy should delay or minimise the development of resistant mould strains. Following either of the above dip treatments, select a benzimidazole group fungicide or SOPP (Sodium orthophenyl phenate) for the packing line treatments.

Option 2. If fruit receives the packing line treatment within 24 hours of harvest.

Use one of the following:

- guazatine;
- imazalil;
- SOPP (see notes below);
- SOPP followed by one of the benzimidazoles or guazatine;

- guazatine followed by one of the benzimidazoles;
- guazatine mixed with one of the benzimidazoles (a special mixing procedure is required [see notes]).

Option 3. If fruit is to be exported treat with at least two approved fungicides, one applied within 24 hours of harvest, either by bulk dipping or by treatment in the packing line.

Approved fungicides:

- guazatine;
- SOPP;
- imazalil;
- benomyl;
- thiabendazole;
- thiophanate-methyl; and
- carbendazim.

Control of various diseases

All of the preceding fungicides or mixtures control green and blue moulds, but guazatine provides only short-term mould protection.

Only guazatine and, to a lesser extent SOPP, control sour rot. Sour rot is caused by a soil fungus that infects fruit through skin injury. It is more severe in wet seasons or on mature fruit and is favoured by 25 to 30°C temperatures.

2,4-D applied in the wax emulsion or incorporated in the final post- harvest fungicide reduces stem-end rot by helping keep the button green. Add a wetting agent.

If mould control is poor, check fungicide rates, application equipment, and times between harvest and treatment. Contact your local extension officer for further advice.

Preparing SOPP dip

SOPP is made by dissolving 2 kg of flake in 100 L of water with enough caustic soda to raise the pH of the dip to 12. This is generally more than 100 g/100 L of dip. Check the dip pH daily to ensure that it remains at pH 12, as acid from the citrus fruit decreases the dip pH. Add caustic soda as required. Top up when necessary with a solution the same strength as the original. **It is essential that SOPP be washed off fruit.** Do not recirculate the rinsing water, and do not expose fruit to SOPP for longer than two minutes as rind injury may result.

Mixing guazatine and benzimidazole

Prepare the benzimidazole fungicide suspension by mixing the benzimidazole powder in a small volume of water, for example 1 mL/g, to produce a smooth paste; then mix the paste with the water in the dipping tank, stirring continuously. Add a non-ionic wetting agent to the dip and stir in well. Finally, add the guazatine and stir in. If the mixture foams excessively during treatment, add two or three drops of anti-foaming agent.

Note: The first treatment must be guazatine or SOPP. As guazatine does not provide long-term protection against mould, it should be combined with either full-strength benomyl, thiabendazole, thiophanate-methyl, carbendazim or imazalil when used as a packing-line treatment.

In addition to the above treatment, lemons and mandarins must be treated with 500 mg/L of 2,4-D either applied in the wax emulsion or incorporated in the final post-harvest fungicide. All fungicide treatments must be applied before waxing. Wettable powder fungicides tend to settle out if allowed to stand, so continuous agitation is needed. Agitation also aerates the dip and prevents build-up of yeasts and bacteria which produce bad odours. When topping up the dips, add the benzimidazole fungicides at twice the strength used to make the original dip. With all other fungicides, top up at normal rates.

Disinfection of sheds, field boxes, graders and brushes

Wash regularly with a hypochlorite-based bleach solution or other disinfectant solution. As these may cause corrosion, wash with clean chlorinated water after treatment.

Centre rot

Alternaria alternata

Symptoms

A dark brown or black stem-end or stylar-end rot of fruit with a black decayed internal core. Often the disease is only evident when the fruit are cut. Affected fruit on the tree may ripen and fall prematurely, but the disease is mainly a post-harvest problem.

Source of infection and spread

Spores of the fungus commonly lodge beneath the 'buttons' of very young fruit but only those which are over-mature or have been weakened by adverse weather or long periods of storage are invaded. The fungus may enter through the stem-end, stylar-end or wounds in the skin.

Navel oranges and Ellendale mandarins are particularly prone to stylar-end infection.

Importance

Only important in Navel oranges and Ellendale mandarins.

Control

1. Use the recommended post-harvest treatment.
2. Avoid storing fruit for long periods and do not store over-mature fruit.

Post-harvest treatment

Refer to blue and green mould for details of the post-harvest treatment.

Sour rot

Geotrichum candidum var. *citri-aurantii*

Symptoms

Sour rot usually occurs as a pale soft decay of ripe fruit in storage. Affected fruit have a putrid smell and larvae of ferment flies are commonly found in the rotting tissues. White fungal growth is often produced on the surface of these fruit.

Source of infection and spread

The fungus is a soil inhabitant and comes into contact with fruit via soil particles which lodge on the fruit or via dips. It infects the fruit through injuries to the skin but in storage, may spread by contact causing 'nests' of rotting fruit. Sour rot is more severe in wet seasons and is favoured by temperatures of 25 to 30°C. Mature fruit, particularly when stored for long periods, is highly susceptible.

Importance

Sour rot is a major post-harvest disease.

Control

1. Use recommended post-harvest fungicides.
2. Handle fruit carefully to avoid damaging the skin.
3. Precool transport fruit at 10°C.
4. Avoid long periods of storage and 'ageing' of late season fruit.

Post-harvest treatment

Refer to blue and green moulds for details of the post-harvest treatment (p. 24).

Stem-end rot

Diaporthe/Diplodia spp.

Symptoms

Stem-end rot is a firm tan-coloured rot beginning at the stem-end and spreading slowly through the fruit. The causal fungi rarely develop visibly on the skin. The margin between healthy and diseased tissues is generally distinct, the decay advancing in streaks corresponding to the fruit segments.

Source of infection and spread

The fungi causing stem-end rot live mainly on dead twigs and branches and the spores spread to fruit during wet weather. After infection, the fungi remain dormant in the 'button' of the fruit until ripening. Infected fruit generally show no external signs of the disease when harvested, although mature fruit left on the tree may develop symptoms.

Importance

Stem-end rot is occasionally serious but control measures are adequate.

Control

1. Spray with the recommended fungicides at 1/2 to 3/4 petal-fall.
2. Use the recommended post-harvest treatment.

Stage: Late September to early October (or when 1/2 to 3/4 of flowers have lost petals).

Chemicals: copper oxychloride or copper hydroxide plus white oil, or miscible summer oil, or narrow range oil.

1. Premix all ingredients separately with a little water.
2. Add each to the spray vat with the agitator working.
3. Add the oil last.

Post-harvest treatment

Refer to blue and green moulds for details of the post-harvest treatment (p. 24).

Custard apples

FIELD DISEASES

Bacteria and mycoplasmas

Bacterial wilt
Pseudomonas solanacearum

Symptoms

Bacterial wilt causes a rapid wilt and death of young trees and a slow decline of old trees. The foliage of seedlings and young trees to three years of age wilts during the warm summer and autumn months, appearing dull green and hanging almost vertically. Growth flushing stops. Leaves usually desiccate within a month without becoming chlorotic and either fall or, in the case of rapid death, remain attached. Darkening of the systemically infected leaf veins can occur but this symptom is more common in rootstock seedlings.

In older trees a slow decline leading to death occurs over two or more years. On affected limbs, mature leaves are shed and the expanding juvenile leaves appear less vigorous, pale green and smaller in size. Branches become sparsely foliated and subject to sunburn. The subsequent spring and summer flushes are sparse and lack vigour but flowering and fruit set may be prolific. However, these fruit remain small and fail to reach a marketable size.

The internal symptoms are easily seen immediately under the bark in the cambium and outer wood tissues of the basal trunk and tap root. The dark, systemic vascular discolouration can be traced from the feeder and primary roots into the tap root and above ground level into the trunk. Dark streaking of the wood is rarely seen in the branches above the graft union. In transverse sections of the trunk at about ground level the outer growth rings are seen to be the most extensively discoloured; crescent-shaped sectors of the inner growth rings are often discoloured from infection in previous years.

The secondary symptoms of collar rot, decay of bark around the crown roots and extensive feeder-root decay become progressively more evident with the demise of the tree.

Source of infection and spread

The bacterium survives in the soil and on a wide range of crop plants and weeds including tomatoes, potatoes, capsicums, eggplants, blackberry nightshade and wild tobacco tree. The bacterium may be introduced in soil or infected rootstocks. Once established, the bacterium can spread rapidly through the crop by irrigation, rainwater and soil adhering to farm machinery and animals.

Importance

The disease is widespread in Queensland's custard apple growing districts. Death of trees of all ages occurs. Young trees may wilt and die quickly following hot wet periods which favour bacterial wilt. Well-established older trees may become unthrifty and sparsely foliated and rot symptoms usually develop after the death of the primary roots.

Control

1. Use resistant cherimoyer rootstocks rather than susceptible sugar apple or segregating atemoya seedlings.
2. Avoid planting into soils known to be infested, if cherimoyer rootstocks are not used.
3. Use disease-free grafted trees propagated in treated potting mix soil.
4. Avoid intercropping custard apple trees with bacterial wilt susceptible annual crops during the orchard establishment phase.
5. Keep weeds, particularly Solanaceous and other bacterial wilt hosts, under control. Maintain a weed-free grass sod between rows and a bare herbicide strip or mulch cover along the row.
6. Avoid waterlogging by ensuring adequate surface drainage, diversion drains and mounding of soils of marginal depth.

Fungi

Armillaria root rot
Armillaria luteobubalina

Symptoms

A slow decline and death of trees. If the base of the trunk and large roots are examined, a cream-coloured fungal growth, sometimes in fans will be seen just beneath the bark. The underlying wood is discoloured and a strong mushroom smell is present. Black threads of the fungus, known as 'shoestrings' may be visible on the surface of the roots.

Honey-coloured mushrooms may form in a clump at the base of an affected tree, particularly in cool wet weather.

Source of infection and spread

The fungus survives in the soil on stumps and roots of various trees for many years. It grows through the soil, contacting and invading living roots.

Importance

Can be serious in some orchards.

Control

Do not plant newly cleared land for at least two years. Deep ripping, removal of tree roots and cultivation of annual crops prior to establishing an orchard reduces the incidence of infection.

Fumigation

Control of Armillaria root rot when replanting infested sites: remove all tree roots by deep ripping and cultivate to a depth of at least 300 mm.

Treat a 3.6 m^2 at each site with methyl bromide (two 680 g cans) released under a 4 m^2 plastic sheet. Ensure edges are embedded to a depth of 150 mm to provide a sealed enclosure. Leave the sheet in place for two days. Remove sheet and aerate the soil for at least two days before planting. Avoid recontamination of the treated area.

Black canker

Phomopsis anonacearum

Symptoms

On fruit, purple spots ranging from small specks to blotches covering half of the fruit. Cracks may form in the affected areas.

Tissues under the spots are brown and rotted to a depth generally no more than 10 mm.

The fungus is also associated with a marginal scorching of leaves.

Source of infection and spread

The fungus survives and produces spores on affected leaves on and under trees; these are spread by wind and rain. Humid conditions favour disease development.

Importance

Minor.

Control

Generally not warranted.

Diplodia fruit rot

Botryodiplodia theobromae

Symptoms

Small black spots expanding rapidly and becoming hard and cracked. External symptoms are very similar to those of black canker. However, internal discolouration and rotting extends well into the fruit, often producing a corky appearance. Affected fruit generally mummify and remain attached to the tree. The fungus also causes a dieback of young twigs.

Source of infection and spread

The fungus survives and produces spores on dead twigs and in the mummified fruit on and under the tree. These are spread by wind and rain. Humid conditions favour disease development.

Importance

Minor.

Control

Remove and destroy mummified fruit and dead branches before fruiting commences.

Purple blotch

Phytophthora palmivora

Symptoms

The first indication of the disease is excessive fall of small immature fruit. Many of these fruit show purple spots. Affected fruit are generally completely brown internally prior to fall.

General

The disease is restricted to fruit on the lower branches and has been found only after heavy rain in trees growing in bare soil.

Importance

Minor.

Control

Generally not warranted. Where serious losses have occurred, mulching of the bare soil under and between the trees is recommended.

Grapes

FIELD DISEASES

Bacteria and mycoplasmas

Leaf and stem spot
Pseudomonas sp.

Symptoms

Leaves: small yellow spots on young leaves which become malformed. The leaf spot may fall out giving the leaves a tattered appearance. Small water-soaked blisters often form on the leaf stalks.

Bunches: individual berry stalks blacken and wither causing malformation of the bunch, berry drop and severe bunch thinning.

Source of infection and spread

The bacterium is a leaf-inhabiting organism which multiplies rapidly under cool wet overcast conditions during early vine growth, and attacks the young leaves and developing bunches.

Importance

Minor unless wet weather is prevalent during flowering.

Control

Copper sprays applied for downy mildew control reduce losses.

Fungi

Armillaria root rot
Armillaria luteobubalina

Symptoms

Vines become yellow, lose vigour, and eventually die. If the crown and large roots are examined, a cream-coloured fungal growth, sometimes in fans, may be seen just beneath the bark. A strong mushroom smell is often detectable. In wet weather, honey-coloured mushrooms with widely separated gills occasionally form at the base of an affected vine.

Source of infection and spread

The fungus survives in the soil on stumps and roots of trees for many years. It grows through the soil invading living roots which it contacts. Once established in a vine, it may spread to other vines by root contact. The disease is worst in damp areas.

Importance

A significant cause of premature death in some districts.

Control

Remove diseased vines with as many roots as possible and burn. Do not replant for at least two years.

Fumigation

Control of Armillaria root rot when replanting infested sites:

Remove as many roots as possible and cultivate to a depth of 300 mm. Treat a 2.5 m^2 at each site with methyl bromide (one 680 g can) under a 3 m^2 plastic sheet with edges imbedded to a depth of 150 mm. Leave the sheet in place for two days.

Bitter rot
Melanconium fuligineum

Symptoms

On berries, large water-soaked areas with a brown centre covered with small black dots, often in concentric rings. The berries become soft and are bitter to taste. On purple-skinned fruit, affected areas may have a pinkish discolouration.

Source of infection and spread

The fungus survives in old mummified fruit left on the vine after harvesting. Spores produced in the small black fruiting bodies are spread by water splash. Infection often follows damage by downy mildew.

Importance

Rare. Generally follows downy mildew infection.

Control

Sprays used for control of downy mildew will reduce losses from bitter rot. The fungicides mancozeb or zineb are preferred when bitter rot has previously been a problem.

Black spot
Elsinoe ampelina

Symptoms

Leaves: young leaves are very susceptible and show small, very dark brown spots, the centres of which become lighter and eventually fall out leaving star-shaped holes. Distortion is common when the disease is serious.

Shoots: small brown spots enlaring to elliptical purple sunken areas with black margins. Growth is stunted and deformed.

Fruit: small circular black spots often with a lighter centre. When fruit stalks are attacked, fruit may not develop.

Source of infection and spread

During winter, the fungus remains inactive inside affected grapevine tissues but in spring and early summer masses of tiny spores are produced.

These wash into the opening buds and subsequently onto healthy leaves, shoots and fruit where infection occurs.

Importance

May be a serious problem especially in wet seasons if protectant fungicides are not applied to early growth.

Control

1. Spray with fungicides especially at budswell and budburst.
2. Remove affected shoots and other debris during winter.

Stage: budswell.

Chemicals: mancozeb, metiram, thiram, ziram.

1. Apply to susceptible cultivars only, for example, Waltham Cross, Purple Cornichon.
2. Ziram is the preferred treatment. However, where dead arm has been a problem previously, use mancozeb.

Stage: budburst.

Chemicals: As for budswell.

1. Apply to all cultivars.
2. Where dead arm has been a problem previously, use mancozeb.

Stage: early shoot growth to flowering.

Chemicals: As for budswell.

1. Apply at fortnightly intervals if disease is present.
2. Mancozeb may be preferred as it also controls downy mildew.
3. Where dead arm has been a problem previously, use mancozeb.

Downy mildew

Plasmopara viticola

Symptoms

Leaves: on young leaves and shoots, pale yellow translucent spots which develop into large, dead areas as the leaves age. In moist weather, a white downy growth develops on the spots on the lower leaf surfaces as well as on affected shoots and tendrils. If dry weather follows, large areas of the leaves turn brown and affected leaves shrivel, die and fall.

Bunches: large purple blotches on fruit stalks and berries which eventually wither and fall.

Source of infection and spread

The white mildew produced in moist weather consists of large numbers of spores which are spread by wind and water to new leaves where they germinate to produce new infections, provided water is present. The spores do not survive in hot dry conditions.

In autumn, the fungus produces overwintering spores (oospores) which survive the winter inside fallen leaves and initiate new infections in the spring.

Importance

Serious and difficult to control in wet seasons particularly in coastal areas. Regular spraying is essential.

Control

1. Spray with fungicides.
2. Remove prunings, fallen leaves and fruit during winter.

Stage: early shoot growth to flowering.

Chemicals: copper oxychloride, mancozeb, zineb, metalaxyl plus copper oxychloride.

1. Apply at fortnightly intervals.
2. Copper preparations are preferred in wet seasons. Do not use copper preparations while vines are flowering.
3. Mancozeb also controls black spot.
4. Metalaxyl plus copper oxychloride is applied at 7 to 21-day intervals. The seven day interval is used when two curative treatments are applied or when prolonged humid and wet conditions occur. Apply a minimum of 500 L/ha of spray before flowering and a minimum of 1000 L/ha after flowering.

Stage: after flowering.

Chemicals: copper oxychloride, mancozeb, zineb, metalaxyl plus copper oxychloride.

1. Apply at fortnightly intervals as required.
2. Do not use copper preparations after berries begin to ripen because of residue problems.
3. Where bitter rot has been a problem previously, mancozeb or zineb are preferred.

Grey mould

Botrytis cinerea

Symptoms

Berries at first show a slight browning which develops into a soft watery decay with grey powdery masses of spores on the surface. In containers, the fungus spreads from fruit to fruit by contact, resulting in nests of rotted fruit.

Source of infection and spread

The fungus is common on dead or dying plant material. Spores, produced in very large numbers, are spread by wind. Cool moist weather favours infection and disease development. If affected berries are packed, extensive breakdown may occur during transit and storage. Damaged fruit are more susceptible to infection.

Importance

Serious but losses can be reduced by careful spraying.

Control

1. Spray with recommended fungicides.
2. Handle fruit carefully to avoid skin damage.

Stage: flowering.

Chemicals: benomyl, carbendazim, iprodione, procymidone, thiophanate- methyl, vinclozolin.

1. Apply at capfall (1 to 10% bloom) and again 10 to 14 days later.
2. These sprays are safe to apply during flowering.
3. It is essential to apply sprays following wet weather.

Phomopsis cane spot
Phomopsis viticola

Symptoms

Leaves: Small dark brown to black spots and yellow haloes. Yellowing and deformation of leaves follows.

Canes: bleached areas containing large numbers of small, black fruiting bodies of the fungus imbedded in the tissues. These areas are often on nodes or on dead stubs of canes. Death of affected canes and arms of the vine occurs over a number of years.

Source of infection and spread

Spores produced in the black fruiting bodies are spread during wet weather and infect young canes, leaves and bunches. The fungus may work back from pruning cuts killing canes or spurs. The disease is favoured by cool, wet weather.

Importance

Occasionally serious in coastal areas.

Control

1. Cut back canes well beyond the visibly affected area and swab large cuts with the recommended fungicide.
2. During dormancy and the growing season, spray vines with the recommended fungicide, mancozeb.

Powdery mildew
Uncinula necator

Symptoms

Leaves: at first, one or more indistinct white patches spreading to eventually cover the leaf with a grey dusty coating.

Canes and leaf stalks: small light brown russetted areas often without the grey dusty mildew.

Fruit: russetting, distortion and cracking often without the grey dusty mildew. The russetting may be difficult to detect on purple-skinned fruit.

Source of infection and spread

The mildew produced on affected areas consists of large numbers of spores which may be carried long distances by wind. The spores may germinate in dry air in contrast to downy mildew where water is essential. High temperatures tend to inhibit development of powdery mildew.

Importance

Uncommon in well-sprayed vineyards. More likely to be seen in dry seasons and in inland areas.

Control

Apply recommended fungicides.

Stage: early shoot growth to flowering.

Chemicals: wettable sulphur or propiconazole.

1. Apply at fortnightly intervals or as required.
2. May be combined with any of the fungicides used for control of downy mildew.
3. Sulphur may cause leaf scorch at temperatures above 24°C.

Stage: after flowering.

Chemicals: As for 'Early shoot growth to flowering'.

1. Apply as required.
2. Fruit that show early symptoms may recover without splitting if thoroughly sprayed.
3. Do not use on wine grapes in the last three weeks before harvest.

Nematodes

Root-knot nematode
Meloidogyne javanica

Symptoms

Galls or swelling on roots. Heavily infested vines are unthrifty and yields are reduced.

General

The problem is most serious in cultivars grown on their own roots.

Importance

Of minor importance where resistant rootstocks are used but can be serious when cultivars are on their own roots.

Control

1. Use rootstocks resistant to root-knot nematodes. Root knot is most serious when cultivars are grown on their own roots.
2. Fumigate the soil before planting.

POST-HARVEST

Rhizopus rot
Rhizopus sp.

Symptoms

A soft watery decay of berries, affected areas becoming covered with a mat of fungal growth containing masses of small black stalked fruiting bodies. The fungus spreads from fruit to fruit by contact causing 'nests' of rotted fruit.

Source of infection and spread

The disease is mainly a post-harvest problem and is rarely seen in the vineyard. The main source of infection is discarded grapes around the packing shed. Initial infection is generally through wounds, bruises, cracks or areas previously affected by downy mildew. However, once established in a case, the fungus may spread rapidly affecting sound fruit by contact.

Importance

Generally a minor post-harvest disease.

Control

1. Handle fruit carefully to avoid skin damage.
2. Do not leave discarded fruit around the packing shed.

Macadamia nut

FIELD DISEASES

Fungi

Anthracnose
Colletotrichum gloeosporioides

Symptoms

On *Macadamia integrifolia* soft brown lesions 10 mm in diameter develop which expand to decay the whole husk. Although decay penetrates to the shell, both the shell and kernel remain sound.

On *Macadamia integrifolia* x *Macadamia tetraphylla* numerous small brown spots, 4 mm in diameter develop on the husk.

Source of infection and spread

Spores are produced on dead leaves, twigs and branches and are spread by splashing water.

Importance

Rare.

Control

The disease has not been serious enough to warrant control measures.

Armillaria root rot
Armillaria luteobubalina

Symptoms

Affected trees may collapse suddenly or decline slowly. Examination of large roots and crown shows black cord-like 'shoestrings' of the fungus on the surface. On the lower trunk and major roots, characteristic white fungal sheets may be visible under the bark. Dark brown to honey-coloured mushrooms may be found at the base of affected trees following wet weather, particularly in winter.

Source of infection and spread

The fungus lives on roots and stumps left after forest-clearing operations. When macadamia trees are planted into these areas, the fungus invades the roots. A small fragment of infected root is sufficient to initiate infection.

Importance

Minor, but can be important on newly cleared land.

Control

1. Remove all tree stumps and roots by deep ripping before planting macadamias.
2. Remove diseased macadamia trees as soon as they are detected. Before replanting the site, remove all roots and cultivate the soil for at least one year.

Fumigation

Control of Armillaria root rot when replanting infested area: remove as many roots as possible and cultivate to a depth of 300 mm. Treat a 3.6 m^2 at each site with methyl bromide (two 680 g cans) under a 4 m^2 plastic sheet with edges imbedded to a depth of 150 mm. Leave the sheet in place for two days.

Macadamia husk spot
Pseudocercospora sp

Symptoms

One to many tan brown circular spots usually 5 to 10 mm diameter occur on the green husk of full size nuts only; spots are not found on twigs, leaves, flowers or young nuts. Husk spots begin as yellow flecks and enlarge to 3 to 6 mm before turning brown in the centre. A pale yellow halo usually surrounds the typically circular 'eyespot' lesion. Under moist conditions mature spots produce a grey velvety carpet of spores. The husk spot is harder to cut through with a sharp blade than the surrounding husk tissue even on fresh green nuts. This characteristic is a useful diagnostic feature.

On the internal surface of the husk, husk spot is evident as a circular tan spot until the nut reaches maturity and the husk turns dark brown. The shell and kernel are not infected by the fungus. Infected nuts may drop from the tree prematurely up to six weeks before maturity is reached; these nuts are very immature, of low oil content and unsuitable for processing.

Source of infection and spread

The husk spot fungus produces wind-dispersed spores on the surface of infected husk tissue, on nuts in the tree or fallen nuts on the ground.

Mechanical harvesting of nuts is more efficient than wind in dispersing the spores through the tree. Young nuts are infected under moist conditions from flowering stage to maturity but symptoms take about 15 to 18 weeks to express, depending on temperature. For the main August to September flowering, infection occurs during spring and summer when showery weather prevails and symptoms are evident from January to harvest date. Nuts set out-of-season from an autumn (April to May) flowering provide spores during spring and summer after the main crop has been harvested in winter.

Importance

Husk spot is the most serious disease of bearing age trees in south east Queensland. Premature nut drop causing up to 50% crop losses has been experienced commercially over several consecutive seasons in the Maleny district. The disease occurs in old orchards of both seedling and grafted cultivars and has not been detected in orchards less than 14 years old. The disease has spread to many coastal orchards between

Caboolture and Nambour and by 1988 had been recorded from Baffle Creek (Bundaberg) and as far south as Gosford, New South Wales.

Control

1. Apply copper fungicides (copper oxychloride or copper hydroxide) instead of copper sulphate in spring to maintain appropriate leaf copper levels. This dual purpose spray applied annually before trees reach bearing age may delay the establishment of epidemics.
2. Manual harvesting of nuts is better than mechanical harvesting as it minimises spore dispersal.
3. Harvest crop as soon as possible after nut maturity is reached and before the main August to September flowering occurs.
4. If husk spot becomes established, harvest crop by midwinter and apply four copper protectant sprays at monthly intervals beginning in early September.
   Rates for high volume spraying - 400 g/100 L.
   Rates for low volume spraying - 8 kg/ha.

Phytophthora trunk canker

Phytophthora cinnamomi

Symptoms

On nursery stock and in young plantation trees, affected plants are stunted, chlorotic and partially defoliated. In mature trees, a discolouration of the bark at the base of the trunk develops, often with gum exudations. If this bark is removed a brown discolouration of the outer wood is visible. As the disease progresses, deep furrowed cankers and irregular areas of dead bark may extend from ground level to the lower branches. Affected plants often set a heavy crop of nuts in the final stages of the disease. Healthy suckers may arise from rootstocks of plants with severely affected scions.

Source of infection and spread

The fungus causing trunk canker is a soil-borne water mould and requires water for spore production and root infection. It invades trees through trunk wounds near ground level. In Queensland, the fungus is often associated with rootlet death even on healthy trees.

The fungus also attacks pineapples, avocados and many native plants.

Importance

An important disease particularly of young trees in the nursery and after planting out.

Control

1. Plant disease-free nursery trees.
2. To minimise waterlogging around the trunk, plant trees with a slight mounding around the trunk rather than in a depression.
3. Avoid wounding trunks during cultural operations.
4. Thorough excision of small cankers followed by application of a wound dressing may control early infections.
5. Spray the lower trunk with the recommended fungicides before the wet season.

Treatment of trunk canker

Cankers must be detected at an early stage. Cut out all discoloured bark and underlying wood and paint the freshly exposed surface with copper oxychloride paste. When callus tissue has started to form, protect with white plastic paint or a tree-sealing compound such as Colgraft®.

In problem areas, spray the lower trunks only with a copper fungicide or metalaxyl plus copper oxychloride before the onset of the wet season.

Mangoes

FIELD DISEASES

Bacteria and mycoplasmas

Bacterial black spot
Xanthomonas campestris pv. *mangiferaeindicae*

Symptoms

Leaves: angular greasy spots delimited by the veins, becoming black and usually raised with yellow margins. Adjacent spots may coalesce, destroying large areas of the leaf.

Fruit: black spots centred on breathing pores (lenticels) with the adjacent tissue having a greasy appearance. The spots enlarge becoming irregular in shape and the central part cracks in the form of a star.

Source of infection and spread

The bacteria on leaves and stems are spread when propagating material is taken from diseased trees. Spread between trees occurs during wet windy weather.

Importance

Bacterial black spot occurs throughout Australia's mango districts and in most countries where susceptible cultivars are grown. Black spot on cv. Kensington is severe in windy areas that lack windbreaks and where trees lack vigour. The disease can be prevalent in low and high rainfall areas. Infected nursery trees have been a major cause of spread of bacterial spot into new orchards.

Control

Pre-harvest control measures are required. The disease can be avoided by using disease-free rootstocks and scionwood. Spray rootstocks with copper oxychloride (4 g/L). Scionwood should be disinfected with 0.35% calcium or sodium hypochlorite solution for five minutes prior to grafting. Do not raise rootstocks or propagate grafted trees under mango trees or close to other inoculum sources.

For field planting, select a site protected from strong winds and provide adequate wind protection around and within the orchard. Resistant cultivars are available (for example Sensation, Kensington Pride (Groszmann selection), Carabao, Nam Dok Mai, Early Gold).

Apply preventative sprays of copper oxychloride 4 g/L. Spray fortnightly during wet weather; monthly at other times. Avoid applications during flowering. Do not market bacterial-spot affected fruit.

Fungi

Anthracnose
Colletotrichum gloeosporioides

Symptoms

Leaves: small black spots which spread and coalesce, forming large dark dry areas. These often crack and fall out.

Flowers: small, black, irregular spots which spread and coalesce to cause shedding and death of flowers.

Fruit: small black specks enlarging to irregular dark brown to black areas. In a moist atmosphere, pink spore masses appear towards the centre of these areas. Surface staining or russetting of fruit may result from spores being washed over the fruit from diseased twigs or flower stalks. Infection of young fruit causes fruit drop.

Source of infection and spread

Spores of the fungus are produced on dead twigs, branches and leaves and are spread by water splash. The fungus may infect green fruit and remain dormant in the tissues until ripening. There is considerable variation in susceptibility of mango cultivars to this disease.

Importance

Major disease but can be controlled by regular spraying and post-harvest dipping.

Control

1. Spray with the recommended fungicides.
2. Remove dead twigs and branches before flowering.
3. Use the recommended post-harvest treatment (see p. 36).

Fungicide sprays

Spray with mancozeb weekly during the blossoming period and then monthly until harvest. During dry weather, blossom sprays may be reduced to fortnightly intervals.

Powdery mildew
Oidium spp.

Symptoms

A white powdery growth on young shoots, flowers and small fruit. Affected fruit fall prematurely. Older fruit show purplish-brown blotchy areas on the skin.

Source of infection and spread

The white mildew consists of large numbers of spores which are dispersed by the wind. The disease is more prevalent in dry weather when the humidity is high.

Mango cultivars vary considerably in their susceptibility to powdery mildew.

Importance

Powdery mildew is a sporadic field problem and is more common during cool weather, particularly when the humidity is high.

Control

Spray with the recommended fungicides.

Fungicide sprays

Under normal conditions, mancozeb applied for anthracnose control will keep powdery mildew in check provided spraying is regular. Apply mancozeb (800 g/kg) at 2 g/L each week during blossoming.

Once powdery mildew becomes well established, mancozeb is not very effective.

POST-HARVEST

Fungi

Anthracnose

Post-harvest control

Dip fruit immediately after harvesting for five minutes in hot water (52°C) to which has been added 1 g/L benomyl (500 g/kg). Temperature must be carefully controlled to prevent fruit damage. With an approximate ratio of 3 L of dip to 1 kg of fruit, no appreciable loss of heat occurs when fruit is added to the tank.

or

Spray fruit for 30 seconds with prochloraz (450 g/L) at 0.55 mL/L. Apply as a non-recirculated overhead spray in water at ambient temperatures. A suitable spray system consists of two TX2 hollow cone nozzles mounted in series above the fruit on the packing line. At an operating pressure of 320 KPa the total delivery rate is 16 L/hr. Two TX2 nozzles are favoured over four TX1 nozzles which deliver the same volume but clog easily.

The rate of decline in concentration of prochloraz has not been defined accurately so it has been approved for use only as a non-recirculated spray. Prochloraz does not control stem-end rot.

Notes:

1. Information on the construction of the dipping tanks is available from QDPI.
2. The required amount of powder should be made into a slurry before adding to the dipping tank. Do not add the powder directly.
3. Agitate before and during dipping by means of a strong recirculating pump as the fungicide settles out or aggregates if the dip is allowed to stand. Agitation is better than stirring or paddling.
4. Top-up as required with a suspension of the same strength as the original dip.
5. The dip should be replaced after three days continuous use or earlier if excessive sap and dirt accumulates. Pre-washing fruit will help prolong dip life.
6. Treated fruit has better skin colour and ripens slightly faster than untreated fruit.
7. Hot benomyl also provides some control of stem-end rot (*Dothiorella dominicana*).
8. Post-harvest treatments will not provide complete disease control. It is important to follow field spray recommendations to reduce the level of post-harvest disease. After harvest, fruit should be allowed to ripen at 20°C. Storage at temperatures above 30°C will increase the likelihood of disease.

Aspergillus fruit rot

Aspergillus spp.

Symptoms

On fruit, water-soaked blotches become light brown and sunken and eventually covered with masses of black spores.

Importance

Minor.

Control

1. Avoid harvesting immature fruit.
2. Observe the hygiene recommendations for control of transit rot.
3. Handle fruit carefully to avoid damaging the skin.

Fungicides

No fungicides are approved for control of Aspergillus rot in mangoes.

Transit soft rot

Rhizopus stolonifer

Symptoms

Small pale spots develop in the peel, spreading rapidly within both peel and flesh. The rotted tissue is soft but generally not discoloured.

Damaged tissue splits readily and under humid conditions, black spore bodies form on white whiskery fungal threads.

Source of infection and spread

The fungus is widely distributed in soil and in decaying organic matter where spores form under moist conditions. It can be present on wood wool used for fruit packing. Spores are dispersed in the air and infect through small wounds. Spore germination occurs in condensate formed on the fruit after removal from cool rooms. Infection may also occur by direct contact with the growing fungus. The disease can spread from fruit to fruit after harvest.

Importance

Minor. It has caused sporadic losses in fruit cooled after harvest, when moisture has condensed on the skin. One or more fruit in the carton may be affected.

The disease may become serious during long-term storage of fruit.

Control

Destroy reject fruit which may harbour the fungus.

Clean the packing shed and equipment with steam cleaning or high pressure hot water. Spray packing equipment and the shed with a sanitising agent.

Wood wool should not be used as a packing material when the fruit are to be cooled as it can act as a source of *Rhizopus* infection under these conditions. It also causes surface scratching.

Fungicides

No fungicides are approved for control of transit rot in mangoes.

Stem-end rot

Dothiorella dominicana, also *Lasiodiplodia theobromae* and *Phomopsis mangiferae.*

Symptoms

A dark brown firm decay starting at the stem-end of the fruit and developing rapidly to involve the whole fruit.

Source of infection and spread

The fungus survives on dead twigs and branches where it produces large numbers of spores. During wet weather, these are spread to adjacent fruit where infection occurs. The rot generally does not develop until the fruit begins to ripen.

Importance

Moderate. May be more important than anthracnose on fruit from drier areas.

Control

1. Avoid harvesting immature fruit.
2. Cool fruit immediately after harvest and store in well-ventilated containers. Post-harvest treatment with hot water and benomyl recommended for anthracnose control is partially effective.

Papaws

FIELD DISEASES

Bacteria and mycoplasmas

Yellow crinkle
Tomato big bud mycoplasma

Symptoms

Leaves: at first, a pronounced yellowing of some old leaves, the stalks of which bend down where they join the trunk. The crown leaves show thin translucent areas along the margins and between the main veins. These areas break away giving the leaves a ragged appearance. The crown leaves eventually become claw-like in appearance and the older leaves dry and fall, leaving the tree a bare pole with a few stunted leaves at the top.

Flowers: abnormal green leaf-like structures develop.

Source of infection and spread

The organism causing this disease infects a large number of other plants and is spread by a small leaf hopper (*Orosius argentatus*). Yellow crinkle is more common during hot dry weather. Young leaves which have not flowered are rarely affected, most losses occuring in plantations during the first fruiting period.

Importance

Periodic epidemics occur.

Control

Maintain a larger number of trees during the first year so that diseased trees can be removed without greatly reducing the stand.

Fungi

Angular leaf spot
Leveillula taurica

Symptoms

Small, scattered patches of white mildew on the underside of the leaf. These patches are characteristically angular in shape as they are delimited by the veins. On the upper side of the leaf, yellowish patches with diffuse margins occur. The disease first affects the older leaves and moves upwards.

Importance

Minor. The disease is rarely seen in Queensland.

Control

Not warranted.

Black spot
Phomopsis caricae-papayae

Symptoms

A black sunken rot of young fruit generally extending from the stem end or a point of contact with a dead leaf. Affected fruit eventually wither and fall. The upper part of the trunk may rot and affected trees may die. However, healthy shoots often arise from lower down on the trunk.

Spotting on mature fruit is also common. A post-harvest stem-end rot may also occur, often resulting in serious losses.

Source of infection and spread

The fungus survives on old fruit and leaf stalks and generally invades trees whose tops have been damaged, for example, by frost.

Importance

Can be serious.

Control

Post-harvest losses may be reduced by harvesting fruit with part of the stalk attached. Cut the stalk level with the shoulders of the fruit.

Brown spot
Corynespora cassiicola

Symptoms

Leaves: light brown circular spots, up to 5 mm wide.

Leaf stalks: long elliptical spots with a dark coating of spores.

Fruit: small dark sunken spots. This symptom is not common.

Source of infection and spread

Spores are produced mainly on the leaf stalks and are spread to adjacent leaves by wind and water.

Importance

Minor.

Control

Control measures generally not warranted.

Powdery mildew
Sphaerotheca humuli sp.

Symptoms

Leaves: light yellow to green patches on young crown leaves, becoming water- soaked and showing a white powdery growth.

Fruit: spreading circular white patches on the surface which may coalesce to cover large areas. As the fruit develops the white mould disappears leaving light grey scarred areas.

Source of infection and spread

The white mildew produced on leaves and fruit contains large numbers of spores which are spread to adjacent trees by wind and rain. The disease is most active during the winter months. It disappears with the spring growth but fruit harvested in midsummer may show scars from early infections.

Importance

Common, sometimes causing serious fruit blemish. Can be controlled with fungicides.

Control

Spray with recommended fungicides.

Fungicide sprays

Spray with wettable sulphur or a mancozeb-sulphur mixture. Dust with sulphur dust to give even coverage. Apply every two to four weeks from May to early October.

Note:

1. Sulphur may cause leaf scorch if applied at temperatures above 24°C.
2. Sulphur dust will have to be reapplied after rain.

Ripe fruit rots

Colletotrichum spp.

Symptoms

At first, brown superficial discolouration of the skin, up to 5 mm wide. These develop into circular, more or less sunken spots up to 30 mm wide. These tend to occur in a group on the outer exposed side of the fruit and often coalesce to form a large rotted area extending well into the flesh.

Source of infection and spread

The fungi causing ripe fruit rots live on dying leaf stalks and produce spores which spread to the fruit, particularly during wet weather. Several of these fungi may infect green fruit, remaining dormant in the tissues until ripening, when they develop rapidly. The environment plays an important part in the incidence of ripe fruit rots. Fruit which have been exposed to the sun or cold winds, as may happen when the older leaves have been killed by winter injury, often ripen prematurely and are particularly prone to spotting. Low temperatures favour disease development by prolonging ripening and giving the fungi more time to develop.

Importance

Cause substantial losses of harvested fruit.

Control

1. Establish plantations in warm sheltered areas.
2. Avoid very susceptible trees when selecting seed.
3. The mancozeb-sulphur mixture applied for pre-harvest control of powdery mildew will also provide some control of ripe fruit rots after harvest.
4. Post-harvest treatment. After harvest spray fruit on the packing line with prochloraz (450 g/kg) at 55 mL/100 L of water at ambient temperature. Spray fruit for two minutes. See mango anthracnose control for details of a suitable application system.

Root rot

Phytophthora and *Pythium* spp.

Symptoms

A yellowing and collapse of older leaves which then hang limply around the trunk before falling. The young crown leaves wilt and the plant may die within a few days, but often trees linger for months in an unthrifty condition with a few small yellow leaves around the crown. These trees are very susceptible to drought and are easily pushed or blown over. Small roots are absent and large ones show a soft wet decay extending towards the trunk. Root rot caused by *Phytophthora palmivora* causes a more rapid decline than that caused by *Pythium* spp.

These fungi may also cause damping off of seedlings and trunk rot of mature trees, the latter generally occuring after wounding. *Phytophthora palmivora* occasionally infects fruit hanging on the tree.

Source of infection and spread

The fungi causing root rot are soil-inhabiting water moulds and require water for spore production and root infection. Severe root rot caused by *Pythium* spp. is usually confined to waterlogged areas of the plantation, and replanting is possible after spelling the land so that papaw residues have decayed. Root rot caused by *Phytophthora palmivora* may occur even in well-drained soils after prolonged periods of wet weather. Replanting of these areas is rarely successful, although with a series of dry seasons the disease may not occur. Before replanting diseased areas, it is essential that the cause of the root rot be accurately determined so that *Phytophthora* infested areas can be avoided.

Importance

Common and often serious, particularly in wet years.

Control

1. Choose only well-drained soils for papaw growing. Construct interception and diversion drains to remove excess water from the plantation.
2. Where Phytophthora root rot has occurred, do not replant with papaws and take care to avoid introduction of the fungus into clean areas.
3. Plant only disease-free nursery trees.
4. To prevent damping off, raise seedlings in sterilised seed beds.
5. To reduce trunk rot, avoid damaging trunks during cultivation.

Seedbed treatments (to prevent damping off): raise seedlings in soil treated with methyl bromide (1 kg/10 m^2 of seedbed) or in potting mixture treated with aerated steam.

Note: before treating soil with methyl bromide, ensure that it is free from undecomposed organic matter.

Undetermined cause

Dieback

Symptoms

At first a bunching and yellowing of the inner crown leaves followed by a slight bending of the stem tip. The crown leaves shrivel and die and often a dark necrotic lesion will be seen where these leaves join the stem. The first leaf to wilt generally has an etched appearance on the undersurface due to a darkening of the smaller veins. The rest of the crown leaves eventually die and the stem begins to die back from the top. The older leaves may yellow but generally remain in position for sometime. If the tree is bearing at the time, fruit may become flaccid or drop and often show severe fungal spotting. Tip bending may be absent and crown death very slow during cold weather.

Source of infection and spread

The cause of this disease is undetermined and little is known about the factors which influence its occurrence. It generally occurs sporadically and severity varies markedly between districts and farms.

Importance

Serious epidemics occur every few years in early summer and autumn, and then often simultaneously throughout a district. The disease is more serious in south Queensland.

Control

There is no prevention at this stage but trees affected when growing vigorously may recover if the stem is cut back below the dead area. This should be done where the leaf scars are close together as at this point solid partitions are present which prevents the entry of rain-water into the otherwise hollow stem, and reduce the likelihood of rotting.

Mosaic

Symptoms

Leaves and stem: a yellowing and stunting of young leaves with translucent areas around the margin. The leaf stalks and upper part of the stem characteristically show narrow, dark green or olive, water-soaked streaks and latex flow is reduced or absent.

Affected plants often produce stunted, multiple side shoots which develop typical mosaic symptoms. The growing points of affected plants are often tunnelled by larvae of the yellow peach moth.

Fruit: young fruit often have light green areas which stand out against the normal dark green colour. These areas are devoid of latex which is still plentiful in the dark green areas.

Source of infection and spread

Unknown. The disease is not caused by papaya mosaic or papaya ringspot viruses.

Importance

Generally minor.

Control

Affected plants are not usually in sufficient numbers to constitute a serious problem but should be destroyed as they are of little commercial value.

POST-HARVEST

Fungi

Ripe fruit rots - see p. 40

Transit rot

Rhizopus stolonifer

Symptoms

Water-soaked spots on the skin of mature fruit spreading rapidly to involve the whole fruit. The flesh becomes soft and watery and affected areas become covered with a white to grey mould with small black stalked fruiting bodies. This may bind the fruit to the side of the carton.

Source of infection and spread

The fungus is common on rotting fruit and invades the fruit through injuries. It occurs more frequently towards the stem-end of the fruit.

Importance

Minor.

Control

1. Handle fruit carefully.
2. Remove all rotting fruit from the vicinity of the packing shed.

Passionfruit

FIELD DISEASES

Fungi

Alternata spot
Alternaria alternata

Symptoms

Leaves: on young leaves, small dead spots occur surrounded by yellow haloes up to 6 mm wide with diffuse margins. On older leaves, dead tissue is surrounded by a yellow or light orange halo.

Fruit: small brown spots surounded by dark green tissue up to 5 mm in diameter. Spots have a greasy appearance.

Source of infection and spread

Spores are produced on infected leaves and fruit and on the dead blossom attached to the fruit stalk. These are readily dispersed by wind and rain.

Importance

The disease is present in most passionfruit crops and can cause severe losses, especially following extended wet periods. Control is difficult with currently available fungicides.

Control

1. Prune vines to remove dead and weak sections of vines. This will allow better spray penetration.
2. Spray with recommended fungicide.

Fungicide sprays

Apply iprodione (100 g/100 L) plus mancozeb (200 g/100 L) during and immediately after extended wet periods. Use mancozeb at 7 to 14-day intervals at other times.

Anthracnose
Glomerella cingulata

Symptoms

As a pre-harvest disease it is most easily recognised on leaves, stems and fruit as tiny brown to black fruiting structures on dead tissue. As a post-harvest disease, it develops following infection of mature fruit in the field. Large areas of the fruit crinkle and collapse, giving fruit a wrinkled appearance.

Source of infection and spread

The fungus is a common inhabitant of dead tissue and is spread by wind and water. Disease development is favoured by warm wet weather.

Importance

Can be serious in north Queensland. Sporadic in south Queensland.

Control

Symptoms develop quickly after harvest. Close inspection of fruit at packing is essential to prevent breakdown during transport.

Brown spot
Alternaria passiflorae

Symptoms

Leaves: large spots up to 10 mm wide, at first chestnut brown in colour, fading to lighter shades of brown. Spots often extend along the main veins.

Stems: usually attacked near leaf axils where brown spots up to 30 mm in length develop. Stems may eventually be cinctured at these points causing death of large sections of vines.

Fruit: light brown circular sunken spots which wrinkle while enlarging and may cover one side of the fruit.

Granadilla (*Passiflora quadrangularis*) shows similar symptoms.

Source of infection and spread

Spores, produced on leaf, stem and fruit spots are blown by wind and rain to adjacent vines. Warm moist weather favours disease development. Susceptible species of wild passion flowers, particularly white passion flower (*Passiflora subpeltata*) and stinking passion flower (*Passiflora foetida*), may be a source of inoculum.

Importance

Can be serious, particularly in wet seasons but can be controlled by a regular spray programme.

Control

1. Spray with recommended fungicides.
2. Prune vines to reduce density and allow better spray penetration.

Fungicide sprays

For control of Septoria spot and brown spot spray with mancozeb or copper oxychloride every 10 to 14 days from October to May and every 21 to 28 days for the remainder of the year.

Fusarium wilt
Fusarium oxysporum f. sp. *passiflorae*

Symptoms

The disease is first noticed as a wilting of one or more shoots and is often followed by total collapse of the plant. If an affected stem is examined, the water-conducting tissues will be discoloured brown or reddish-brown.

Source of infection and spread

The fungus is a soil inhabitant and may survive there for many years as dormant resting spores. Infection occurs through the water-conducting tissues of plant roots.

Importance

Minor. Adequately controlled by the use of resistant rootstocks.

Control

Grow cultivars grafted onto wilt-resistant seedlings of the golden passionfruit (*Passiflora edulis* f. *flavicarpa*). A seed line highly resistant to Fusarium wilt is available from QDPI.

Phytophthora blight

Phytophthora nicotianae var. *parasitica*

Symptoms

Leaves: blackening and death of young tip growth. Large water-soaked areas on leaves which become light tan in colour. Affected leaves fall readily and vines may be defoliated.

Stems: areas are at first purple and later turn brown above the graft union. These may completely girdle the stem causing wilting and collapse of the vine.

Fruit: large grey-green water-soaked areas. Affected fruits fall readily and, in wet weather, become covered with white fungal growth.

Source of infection and spread

Spores are initially produced in wet soil beneath vines and splashed onto the lower leaves. Further spread occurs during wet, windy weather.

The fungus is widespread and affects many other crops including pineapple, strawberry, citrus, tomato, and tobacco.

Importance

Can be devastating in exceptionally wet periods.

Control

1. Spray with recommended fungicides.
2. Prune vines to reduce density and allow better spray penetration.
3. Develop a grass sward under the vines.

Fungicide sprays

If Phytophthora blight has been a problem in previous years, use copper oxychloride in preference to mancozeb when prolonged wet weather occurs. Reduce the spraying interval to 7 to 10 days.

Note: Copper oxychloride at rates higher than recommended leave unsightly spray residues on the fruit.

Scab

Cladosporium herbarum

Symptoms

Small circular translucent spots develop on leaves and young shoots, becoming covered with grey powdery masses of spores. Affected leaves fall readily. Similar spots may form on fruit with the tissues beneath growing into hard raised scabs up to 3 mm wide. Fruit are disfigured and unsuitable for fresh fruit marketing.

General

The disease can be a problem in cooler high altitude areas.

Importance

Minor.

Control

Not warranted.

Septoria spot

Septoria passiflorae

Symptoms

Leaves: brown spots up to 2 mm wide with minute, black dots (fruiting bodies) scattered over the surface. Infected leaves fall readily and vines may be defoliated.

Stems: similar spots to those on leaves but rarely seen.

Fruit: light brown blotches studded with minute black fruiting bodies, often coalescing to cover large areas of the fruit. Affected fruit ripens unevenly.

Granadilla (*Passiflora quadrangularis*) shows similar symptoms.

Source of infection and spread

Spores produced in the black fruiting bodies are blown to adjacent vines during wet windy weather. Warm moist weather favours disease development.

Importance

Can be serious, particularly in wet seasons but can be controlled by a regular spray programme.

Control

1. Spray with recommended fungicides.
2. Prune vines to reduce density and allow better spray penetration.

Fungicide sprays

For control of Septoria spot and brown spot spray with mancozeb or copper oxychloride every 10 to 14 days from October to May and every 21 to 28 days for the remainder of the year.

Undetermined cause

Base rot

Symptoms

A soft decay of the base of the stem and the upper part of some of the main roots. Disease develoment is generally slow so that an affected plant may survive for months. If the vine is in a vigorous condition, new tissue will often be formed on the outside of the stem base as fast as it decays on the inside with the result that it becomes hollow and several times normal size. In this case, the vine may tolerate the disease.

Importance

Base rot can cause considerable losses on some farms.

Control

Injuries to the base of the plant predispose it to base rot. Do not damage the base of the stem or roots during cultivation.

Viruses

Woodiness

Passionfruit woodiness virus

Symptoms

Leaves: light and dark green mosaic patten often with a light yellow speckle. Leaves may be puckered and crinkled.

Fruit: small and misshapen with an abnormal thick hard rind and small pulp cavity.

Granadilla (*Passiflora quadrangularis*) shows similar symptoms.

Source of infection and spread

The virus is transmitted from infected to healthy vines by aphids feeding for short periods while migrating through the area. This virus may also be spread in infected scion wood and on pruning implements.

Most species of wild passion flowers are susceptible and are a source of the virus.

The purple passionfruit (*Passiflora edulis*) invariably contracts the disease and may show extensive fruit deformation. Once infected, vines cannot be cured and are worthless for commercial production.

Importance

Major. Several strains of the virus occur in Queensland.

The 'common' strain is probably comprised of a range of isolates which produce severe woodiness symptoms on the purple passionfruit but are tolerated by hybrids (*P. edulis* Sims x *P. edulis* f. *flavicarpa*). In these hybrids a range of mild to severe foliage mosaic symptoms develop but few woody fruit.

A 'mild strain' which has less effect on both purple passionfruit and hybrids has been tested experimentally.

The 'tip blight' strain is selectively derived from the corky passion vine (*P. suberosa*) weed source and causes necrotic terminal blighting of *P. edulis*. A new severe woodiness virus strain causes severe leaf mosaic and woody fruit even in hybrids. It often causes a necrotic speckle on older leaves. This strain was propagated and spread in nursery material, particularly in the 'purple gold' selection.

Control

The severe strain can be controlled by careful selection of tip sources from areas where this strain is not common. Use of tolerant hybrid material has provided reasonable control of woodiness disease. However, contamination by the severe strain, spells of cool weather or otherwise unfavourable environmental conditions can also cause an increased incidence of woody fruit. Plant on good aspects in warm situations. The 3-1 hybrid is more susceptible to loss by the severe strain.

Pears

FIELD DISEASES

Fungi

Bitter rot

Glomerella cingulata

Symptoms

The disease affects fruit only. Small circular light brown, slightly sunken spots enlarging rapidly to rot large areas of the fruit. Raised pink spore masses, often in concentric rings, appear near the centre of each spot. The spots generally appear when the fruit approach maturity and are soft but not distinctly mushy or leathery.

Source of infection and spread

The fungus overwinters in fruit 'mummies' from the previous season and in bark crevices. Spores are released in warm humid weather to infect fruit. The pink spore masses which then develop on fruit provide large numbers of spores allowing rapid spread of the disease. As ripe fruit is most susceptible, the disease seldom appears until late in the season.

Importance

Summer disease which may be serious as fruit nears maturity. It can be controlled by spraying and attention to orchard hygiene.

Control

1. Remove and destroy all mummified fruit at pruning and diseased fruit during the season.
2. Spray with fungicides.

Stage: November to maturity.

Chemicals: mancozeb, zineb, maneb + zineb, ziram.

1. Spray at 21-day intervals with recommended fungicides from November to near harvest.
2. Remove and bury fruit affected by bitter rot.

Black spot

Venturia pirina

Symptoms

Leaves: olive green to dark brown or black spots particularly on the undersurface. On young leaves, spots tend to spread along veins. When the disease is severe, leaves are small, distorted and may fall.

Twigs: irregular black or purple patches occur when the disease is severe.

Fruit: small dark spots enlarging to become brown and corky with a black, broken margin. Fruit become distorted and cracked.

Source of infection and spread

During winter, the fungus grows within old diseased leaves on the ground beneath the trees. Here, the fungus forms minute fruiting bodies (perithecia) in which large numbers of primary spores (ascospores) are produced.

With spring rains, these are forcibly ejected into the air and carried by wind and rain to developing leaves and flowers where infection occurs. By midsummer, almost all primary spores have been released.

Two weeks after infection, spots appear on the leaves and fruit. On the surface of these spots, secondary spores (conidia) are produced and are spread to adjacent trees by wind and rain. Showery weather favours new infections. Secondary spores are produced until autumn leaf fall.

Importance

Can be serious, particularly in wet seasons. Controlled by a regular spray programme combined with eradicant sprays after black spot warnings if rain interrupts regular spraying with protectant fungicides.

Control

1. Spray with fungicides, especially from green tip to 95% petal fall.
2. Apply the recommended pre-leaf fall spray.

Stage: green tip

Chemicals: copper oxychloride, lime sulphur plus wetting agent.

Green tip occurs earlier in pears than in apples.

Stage: blossoming

(Blossom break to petal-fall). Chemicals: protectant fungicides - dithianon, mancozeb, metiram, thiram, ziram, maneb + zineb. Eradicant fungicides - dodine powder or liquid, fenarimol.

1. Apply protectant fungicides at blossom break, full bloom and petal-fall.
2. If these cannot be applied because of rain, spray with an eradicant fungicide according to black spot warnings. Apply dodine within two days or fenarimol within four days of warning.
3. Some products carry warnings about application within certain times of oil sprays. Check label for details.
4. Dodine may russet fruit of Williams Bon Chretien.

Stage: petal-fall to harvest.

Chemicals: as for 'blossoming'.

1. Spray every 14 to 21 days. Use the shorter interval if wet weather occurs.
2. If bitter rot has been a problem previously, use mancozeb or zineb in preference to the others.

Stage: April to May.

Chemical: urea (5 kg/100 L).

Spray at 10% leaf-fall to prevent overwintering of the fungus.

Stage: dormancy.
Chemicals: urea, paraquat.
Ground treatment after leaf-fall.

Brown rot
Sclerotinia fructicola

Details are the same as those given for brown rot in apples (p. 2).

Importance

Disease of mature fruit of minor importance in well-sprayed orchards.

Control

Details are the same as those given for brown rot in apples.

Fabraea scald
Fabraea maculata

Symptoms

Leaves: small purplish slightly raised spots increasing in size to about 5 mm wide and changing in colour to dark purple and then black.

Fruit: small reddish spots gradually darkening to resemble those of black spot. Severe cracking may occur on young fruit.

Importance

Rare on pears.

Control

The spray programme recommended for black spot controls this disease (p. 47).

Nematodes

Root-lesion nematode
Pratylenchus penetrans

Details are the same as those given for replant failure in apples (p. 5).

Importance

Can be serious where preplant fumigation is not practised.

Control

Details are the same as those given for replant failure in apples (p. 5).

POST-HARVEST

Fungi

Blue mould
Penicillium expansum

Symptoms

Fruit at first shows soft pale brown watery spots with the decayed portions sharply separated from the healthy tissues. Spots enlarge rapidly under favourable conditions and may completely envelop the fruit. Under warm moist conditions, at first white then pale blue and finally blue-green powdery clumps develop on the surface of the spots. The fruit eventually becomes a soft watery mass with a characteristic musty smell.

Source of infection and spread

Blue mould is principally a post-harvest disease and is rarely found in the orchard. The most important source of the fungus is diseased and discarded fruit in and around the packing shed. The fungus generally invades the fruit through wounds in the skin. Once established, it develops rapidly under warm moist conditions and produces vast numbers of spores. However, in dry cold air there is little or no surface mould even when the fruit is completely decayed.

In containers, the fungus may spread from fruit to fruit by contact, resulting in 'nests' of mouldy fruit. Also, fruit weakened by long periods of storage or poor storage conditions may be invaded through the pores in the skin resulting in multiple infections known as 'spot waste'.

Importance

Most important post-harvest disease. Can be controlled by post-harvest dipping.

Control

1. Handle fruit carefully to avoid damaging the skin.
2. Keep packing sheds clean and free from rotting fruit. Wash graders regularly.
3. Cool fruit promptly to restrict disease development.
4. Reduce dust in packing shed areas by watering roadways and entry points regularly.

Post-harvest fungicide dip (for control of blue mould and grey mould). Dip in benomyl, imazalil, iprodione or thiophanate-methyl for 30 seconds.

A mixture of two of these fungicides is generally used.

Note:
1. The required amount of fungicide should first be made into a slurry before adding to the dipping tank. Do not add the powder directly to the tank.
2. Agitate while dipping to keep the fungicide in suspension.
3. Top-up regularly with a suspension of the fungicide equivalent in strength to that of the original dip.

4. The fungicide may be included in diphenylamine anti-scald dips.
5. Thiophanate-methyl is registered for the control of blue mould in Packham pears only.

Disinfectant for packing and grading equipment
Wash regularly with a hypochlorite-based bleach solution used at household rates for example 150 mL of 3.5% sodium hypochlorite/L water. As this corrodes some materials, wash thoroughly with clean water after treatment.

Grey mould

Botrytis cinerea

Symptoms

Fruit only affected. A soft brown rot develops which eventually envelops the fruit. Rotted fruit has a pleasant fermented odour. In humid conditions, patches of ashy grey powdery spores develop on the surface of affected fruit. In advanced stages of decay, small black hard fruiting structures of the fungus may be present. The fungus may spread to other fruit in the container causing 'nests' of rotted fruit.

Source of infection and spread

Grey mould is principally a post-harvest disease, although infection of fruit on the tree may occur, particularly if cool wet weather occurs around harvesting. Infection occurs mainly through skin injuries.

Importance

A disease of mature fruit of minor importance in well sprayed orchards.

Control

1. Handle fruit carefully to avoid damaging the skin.
2. Keep the packing shed free from rotting fruit.
3. Cool fruit promptly to restrict disease development.

Post-harvest fungicide dip as for control of blue mould (p. 48).

Pineapples

FIELD DISEASES

Fungi

Base rot and white leaf spot
Ceratocystis paradoxa

Symptoms

Base rot: a black rot of the butt of planting material. Softer tissues are destroyed and stringy fibres remain. If affected material is planted, butt decay is followed by wilting of leaves.

White leaf spot: yellow to brown spots, several centimetres in length, later drying to become papery and straw-coloured.

Source of infection and spread

The fungus is widespread and is important in the breakdown of pineapple residues. It enters only through wounds and is most active in warm wet weather.

Base rot: the fungus enters most commonly through the unhealed scar where the planting material has been removed from the parent plant. These infections often begin to develop while material is being stored and may not be obvious at planting.

White leaf spot: occurs in autumn during wet windy weather.

Importance

Base rot: serious if tops are not cured or treated with fungicide.

White leaf spot: common but of no economic importance.

Control

Base rot:
1. Do not leave a portion of fruit attached to the top when picking.
2. Do not plant fresh tops unless they are first treated with the recommended fungicide benomyl.
3. Store planting material in a single layer with the butts exposed to the sun. If prolonged wet weather occurs, spray upturned butts with the recommended fungicide.
4. Improve soil drainage and avoid planting during wet weather.

White leaf spot: control measures are not warranted.

Fruitlet core rot
Fusarium moniliforme

Symptoms

External: one or more 'eyes' may fail to colour, a condition referred to as 'green eye'.

Internal: a light to dark brown round or triangular soft rot of the fruitlet cores.

Source of infection and spread

The fungi colonise decaying flower remnants in the cavities under the eyes and then enter the fruit through stylar canals, nectary ducts, insect injuries and internal growth cracks. Infection may occur from flowering to fruit maturity. The disease is often severe when wet weather occurs within six to eight weeks of harvest after the fruit have developed in a relatively dry season.

It is most severe in plant crop fruit and is generally more severe in May and June. However, serious losses may occur in March and April in cultivars of the Queen group (Ripley Queen and MacGregor), which are particularly susceptible.

Importance

Serious in rough-leafed cultivars.

Control

With particularly susceptible cultivars, control the crop so that it does not flower in December.

Green fruit rot
Phytophthora cinnamomi

Symptoms

A general, water-soaked rot of green fruit.

Source of infection and spread

The fungus is a soil inhabitant and requires water for spore production and fruit infection. Spores produced in surface water are splashed onto fruit near the ground. Only ratoon crops are affected. Serious losses generally follow heavy rains when plants have lodged as a result of root rot.

Importance

Generally minor but serious losses may occur in ratoon crops following heavy rain.

Control

Control as for top and root rot (p. 52).

Top and root rot
Phytophthora spp.

Symptoms

Top (heart) rot: Firstly, colour change of the heart leaves to yellow or light brown with a red tinge. The leaf edges curve back and leaves are easily pulled from the plant. Internally, the soft stem and leaf bases are rotted and smell unpleasant. The growing point of the stem has a cheese-like appearance.

Root rot: Firstly, the leaves change colour similar to top rot symptoms. However, the outer leaves become limp and die back from the tips. By this stage, the root system has collapsed and plants are easily pulled from the ground. Fruit from diseased plants are small and unmarketable.

An infection through the root tip often extends into the stem to induce top rot.

Source of infection and spread

The fungi are soil inhabitants and require water for spore production and root infection. Serious losses generally follow heavy autumn and winter rains. Tops and slips have soft stem tissues easily invaded by the fungi and, particularly when planted in autumn, are more susceptible to rot than suckers. Also, the long stem of the sucker means that more tissue has to be invaded before symptoms appear.

Losses from top and root rot are more severe on poorly drained soils. However, tops may be affected even on relatively well-drained soils if prolonged wet weather follows planting.

Phytophthora cinnamomi causes heaviest losses in autumn and winter in all pineapple-growing districts of Queensland, whereas *Phytophthora nicotianae* var. *parasitica* is responsible for spring and early summer losses in the Mary Valley and autumn and winter losses in central and north Queensland.

Importance

Most serious diseases of the crop.

Control

1. Use well-drained soils for pineapple growing.
2. Plant on raised beds at least 230 mm high after settling.
3. Construct drains to intercept runoff before it reaches the plantation.
4. Within the plantation, divert water into inter-row drains so that it can be removed as rapidly as possible without erosion.
5. Deep rip down the slope before hilling if subsurface soil compaction is evident.
6. Mulch high beds with polythene. This sheds excess water into the inter-row drains, reduces bed erosion, encourages surface rooting and keeps soils warmer. However, do not mulch beds less than 230 mm high because, in low beds and on flat ground, the roots are often in waterlogged soil following wet weather and the polythene mulch retards drying. This prolongs conditions suitable for development of the fungi.
7. Apply the recommended fungicides.
8. Before planting, apply sulphur to reduce soil pH.

Foliar drenching with fungicides (for control of top and root rot). Apply either metalaxyl or fosetyl. Apply the first treatment immediately after planting, preferably as a hand drench, and repeat the applications by spraying metalaxyl or fosetyl at four to eight-week intervals. Use the shorter interval during periods of wet weather or in areas with a past history of the disease. Use 2000 L/ha for young plants increasing to 5000 L/ha for large plants.

Note:

1. Do not apply metalaxyl within one month of harvest or fosetyl within seven days of harvest.
2. Do not use a wetting agent with these fungicides.

Nematodes

Root-knot nematodes (*Meloidogyne* spp.) and the root-lesion nematode (*Pratylenchus brachyurus*).

Symptoms

Root-knot nematodes produce distinct terminal swellings on the roots. The root-lesion nematode produces brown areas which may girdle roots and cause their premature death.

Importance

Serious in most fields.

Control

Monitor nematode populations in each field and use the data to make decisions on control measures. Key sampling periods are at the end of the ratoon crop and 6 to 15 months after planting.

Fumigate the soil before planting. Apply a post-plant nematicide in fields with a history of nematode problems; in areas where high nematode populations were observed in the previous ratoon crop; or in fields where root-knot nematode build-up is detected 6 to 15 months after planting.

Preplant fumigation. Inject ethylene dibromide (EDB) at recommended label rates to a depth of 200 to 250 mm into formed beds in lines directly beneath each intended plant row. From each tine, apply 4.3 mL of diluted product of EDB193/m of row. Before fumigation, the soil should be cultivated deeply and be free from undecomposed residues and surface clods, slightly below field capacity and above 15°C. Rolling after fumigation is desirable. Allow at least two weeks before planting, particularly if the soil is cold or wet.

Post-plant nematicide treatment. Apply fenamiphos to the ratoon crop as two sprays, six weeks apart, each at 12 L/ha.

Undetermined cause

Mealybug wilt

Symptoms

Leaves die back from the tip and often change colour to pale green or yellow. The inner heart leaves generally remain normal. Although roots die, they do not degenerate to the same extent as when affected by Phytophthora root rots. Plants often recover and serve as sources of the disease

Source of infection and spread

The disease is probably introduced in planting material which may not show obvious disease symptoms. Once established, it is spread by the mealy bug (*Dysmicoccus brevipes*) a sedentary insect, which is moved from plant to plant by attendant ants. Warm weather favours the build-up of mealy bugs and this is when most serious outbreaks of the disease occur.

Importance

Occasionally serious.

Control

1. Use planting material from wilt-free areas.
2. Use recommended insecticides for mealybug and ant control.
3. Eradicate badly affected areas.

Insecticide for control of mealybugs. Spray with diazinon at recommended label rates plus white oil or miscible summer oil or narrow range oil added at a rate of 1 L/100 L. Use a minimum of 3000 L/ha of spray. Apply when pests are first seen and repeat at 90-day intervals or as necessary.

POST-HARVEST

Bacteria and mycoplasmas

Marbling and pink disease

Erwinia ananas (marbling) and undetermined bacteria (pink disease).

Symptoms

Only fruit are affected. There are no external symptoms of either disease. Internally, with pink disease, the flesh may be water-soaked or light pink and have an aromatic odour but none of these symptoms may be immediately obvious. However, during canning, affected tissue turns dark brown.

Importance

Both diseases can be minor problems during canning of the fruit.

Control

None available.

Fungi

Water blister

Ceratocystis paradoxa

Symptoms

A soft watery rot of the flesh of the fruit with the overlying skin glassy water-soaked and brittle. Eventually, the skin, flesh and core disintegrate and the fruit weeps, finally leaving an empty fruit carcass containing a few black vascular fibres.

Source of infection and spread

The fungus is widespread and is important in the breakdown of pineapple residues. It enters only through wounds and is most active in warm wet weather. Infection occurs through bruises and growth cracks but mainly through the broken fruit stalks. The disease is most severe from Januray to April when the summer crop is harvested.

Importance

The major post-harvest disease.

Control

1. Remove pineapple refuse from in and around the packing shed. Treat the the shed with the recommended disinfectant.
2. Handle fruit carefully to avoid bruising.
3. Reject sunburnt and damaged fruit
4. Dip the base of the fruit in the recommended fungicide within five hours of harvesting.

Post-harvest treatment (for control in fruit for the fresh market). Dip the base of the fruit in benomyl (480 g/100 L) within five hours of picking. This treatment is more important for fruit harvested in late summer and early autumn, during wet weather or where it is being consigned to distant markets.

Shed disinfectant. Remove reject fruit and pineapple trash from the shed daily. Clean and sanitise the shed at least once per week. Spray the shed with a sanitising agent.

(a) Chlorine solution: use 200 mL/100 L of a sodium or calcium hypochlorite preparation containing 100 g/L (10%) available chlorine.
(b) Quaternary Ammonium Compounds, for example Applied 3-300®. Use 200 mL/100 L. This product has a residual effect. Apply only to clean surface. Avoid inhalation.
(c) Formalin: use 2 to 5 L/100 L of a 40% solution. **Caution:** This a potent product which is unpleasant to use and should not be inhaled. In confined spaces wear a full face mask with the correct canister. Wear protective clothing.

Yeasty rot

Saccharomyces spp.

Symptoms

The first symptom is a bubbling exudation of gas and juice through the crack or injury where infection occurred. The skin turns brown and leathery and, as the juice escapes, the fruit become spongy. Internally, the decaying flesh is bright yellow and contains large gas cavities.

Source of infection and spread

The fungus infects through growth cracks, injuries and sunburnt areas. Growth cracks commonly form when rain follows a long dry period. Only ripe fruit are affected and high temperatures are necessary for disese development. The disease may occur in the plantation or as a post-harvest problem.

Importance

Minor.

Control

1. Protect plants from sunburn.
2. Take care not to damage fruit when handling.

Stone fruit

FIELD DISEASES

Bacteria and mycoplasmas

Bacterial canker

Pseudomonas syringae pv. *syringae*

Symptoms

Leaves: brown irregular spots which separate from the surrounding tissue and drop out giving the leaf a shot-hole appearance.

Twigs: withering of young growth resulting from the formation of sunken spots called cankers. Gumming is often present on these cankers.

Limbs: circular to elongated cankers with the underlying tissue discoloured brown. Gumming is common.

Source of infection and spread

The bacterium overwinters in the cankers on twigs and limbs. In spring, large numbers of bacterial cells are produced and spread during wet windy weather. Twig infection commonly occurs through leaf scars in the autumn. The bacteria may be spread in infected buds used for grafting.

Importance

The disease has been recorded on apricots and plums in Queensland. The distribution is restricted but is sometimes severe, particularly on apricots. Very difficult to control.

Control

Spray with the recommended copper fungicides at early leaf fall and again at bud movement and pre-bloom.

Stagc: carly bud movement.
Chemical: copper oxychloride.
Stage: 7 to 10 days later (pink stage in apricots).
Chemical: copper oxychloride.

1. Not necessary on European plums.
2. Apricots need only be sprayed if the disease is a problem. However, if freckle has been a problem previously, spray with copper oxychloride, mancozeb or thiram.

Stage: blossoming.
Chemical: copper oxychloride.

1. Apply just before full bloom.
2. Necessary only on very susceptible Japanese plums such as Doris, Shiro, October Purple and Mariposa.
3. Lower rate of chemical is recommended because of a risk of phytotoxicity.

Stage: 30 to 50% leaf-fall.
Chemical: copper oxychloride.
Stage: late leaf-fall - for plums.
Chemical: copper oxychloride.

1. Apply one or more sprays during winter. Aim to keep trees 'blue' with copper fungicide during winter. Remove severely affected trees.
2. Necessary only where bacterial canker is a problem.

Bacterial spot

Xanthomonas campestris pv. *pruni*

Symptoms

Leaves: small, greenish-yellow spots which, in wet weather, enlarge into angular water-soaked areas. These dry out, become dark brown to purple, eventually separate from the surrounding tissue and drop out giving the leaf a shot-hole appearance. If the spots are close together and in large numbers, the leaves turn yellow and may fall. In peach trees, spotting is frequently most severe towards the leaf tips.

Twigs: greasy dark green elongated areas developing into tan sunken areas called cankers.

Fruit: on peach tan, pinpoint spots occur which crack to form pits on the surface. Cracks and pits may be quite extensive and are often associated with a gummy material. On ripe fruit, a green halo may surround the pits.

On plums, circular greasy spots develop which darken and crack in the centre as the fruit enlarges.

Source of infection and spread

The bacterium overwinters in a dormant state in twig cankers and in leaf buds. In spring, large numbers of bacterial cells are produced and are spread during wet windy weather. Twig infection mainly occurs at leaf fall through the leaf scars. The bacteria may also be spread in infected buds used for grafting.

Importance

A major disease in susceptible cultivars such as Doris and Santa Rosa plums. More serious and difficult to control in wet seasons. Plums and Starking Delicious peaches are most severely affected.

Control

Spray with the recommended copper fungicides at early leaf-fall and again at bud movement and pre-bloom. See bacterial canker control section (p. 55).

Crown gall

Agrobacterium tumefaciens

Symptoms

Hard woody irregularly shaped galls up to 100 mm in diameter and with rough surfaces form at the crown of young seedlings. Smaller galls up to 40 mm in diameter occasionally form on secondary roots. Sometimes galls form above ground on young stems and rarely on the lower branches. Vigour of trees is reduced and death may result.

Source of infection and spread

The bacterium survives in nursery and orchard soils and is spread by water to the roots of young seedlings.

Importance

The disease can cause significant losses in peaches and occasionally affects plum and cherry. A means of biological control is now available.

Control

1. Before germination dip seed in recommended bacterial inoculant.
2. Remove and burn affected nursery trees.
3. Dip nursery trees in recommended bacterial inoculant at time of planting.

Treatment of planting material

Dip in Nogall® powder (a preparation containing *Agrobacterium radiobacter*). Prepare the dip by mixing a 250 g packet of Nogall in 12 L of water in a clean plastic bucket or garbage can. Stir vigorously to ensure thorough mixing.

Seeds. Before planting in the nursery, dip seeds in the mixture, making sure they are all thoroughly wetted. Plant immediately after treatment. Do not allow the seed to dry.

Seedlings. Dip and thoroughly wet the roots within two hours of lifting. Trim damaged roots before dipping. Do not allow treated roots to be damaged or exposed to the sun for a long period. Dip the trees again before planting in the orchard.

Cuttings. When the cuttings are used for propagation, dip before applying rooting compounds.

Note:

1. Use the dip within 48 hours of preparation.
2. Nogall® may be stored for up to 6 months at 4°C but must not be frozen.
3. Nogall is available from Root Nodule Pty, PO Box 367, WOYWOY, NSW 2256.

Fungi

Armillaria root rot

Armillaria luteobubalina

Symptoms

A decline and death of trees generally in groups. A cream-coloured fungal growth, sometimes in fans, is seen just beneath the bark of the crown and large roots. The underlying wood is discoloured and a strong mushroom smell is often detectable. Black threads of the fungus, known as 'shoe-strings', may be visible on the surface of the roots. In wet cold weather, honey-coloured mushrooms with widely separated gills occasionally form at the base of an affected tree.

Source of infection and spread

The fungus may survive in the soil on stumps and roots of trees for many years. It grows through the soil, invading living roots with which it comes into contact. Once established in a tree, it may spread to neighbouring trees by root contact.

Importance

Causes significant losses of mature trees.

Control

Remove diseased trees and as many roots as possible. Do not replant for at least two years.

Fumigation

When replanting infested sites, first remove tree roots thoroughly and cultivate to a depth of 300 mm. Treat 3.6 m^2 at each site with methyl bromide (two 680 g cans) under a 4 m^2 plastic sheet with edges embedded to a depth of 150 mm. Leave the sheet in place for two days.

Caution: methyl bromide is a highly toxic gas - use with extreme care according to label directions.

Brown rot

Sclerotinia fructicola

Symptoms

Blossom blight: affected blossoms turn brown, wilt and the petals wrap around the centre of the flower. Gum may exude from the flower stalk.

Twig blight: a twig bearing affected blossoms may develop light brown spots up to 30 mm long, often with gum. In moist weather, masses of light brown spores form on the blossoms and spots particularly at the edge of the gum. Leaves on affected twigs wilt, turn brown and fall and the twigs begin to die back. In apricots, the green fruitlets may also be attacked and develop brown rot.

Fruit rot: small round brown spots enlarging rapidly in moist conditions to cause a firm brown rot eventually involving the whole fruit. Light brown powdery clumps of spores, often arranged in concentric rings, develop on the rotted areas. If affected fruit are allowed to remain in the orchard on either the tree or the ground, they shrivel up into hard bodies called 'mummies'.

Source of infection and spread

The fungus survives the winter in 'mummies' and in diseased fruit stalks and twigs. In spring, it becomes active again, producing large numbers of spores which are spread to the emerging blossoms. If these remain wet for more than 10 hours, infection occurs and blossom blight develops. Twig blight may follow.

Young fruit may be infected by spores from these twigs but symptoms only develop when the fruit mature. Maturing fruit may be infected at any time during warm wet weather, spores being spread by wind, rain and insects such as dried fruit beetles (*Carpophilus* spp). The fungus may penetrate the broken skins of fruit but insect injuries, wind and hail damage and bruising greatly increase infection.

Post-harvest contamination of fruit may occur during handling and packing, resulting in extensive breakdown in transit and storage.

Importance

A major disease which is particularly serious and difficult to control in wet seasons. Requires timely spraying and crop hygiene for control. Post-harvest phase controlled by dipping. Apples, pears and quinces are also occasionally affected.

Control

1. Spray trees with recommended fungicides during blossoming and the month before harvesting.
2. Use the recommended post-harvest fungicide treatment.
3. Prune out and destroy all diseased twigs and 'mummies' during winter.
4. During the season, remove and bury rotten fruit on the tree or on the ground.
5. Handle fruit carefully to avoid damaging the skin.
6. Observe hygiene in the packing shed, cleaning equipment regularly with disinfectant and removing waste fruit daily.

Stage: blossoming (blossom blight phase).

Chemicals: two sprays at early (1 to 10%) and full bloom with iprodione.

If spray schedule is interrupted, apply one spray of benomyl, carbendazim, procymidone, or vinclozolin at full bloom.

1. Necessary on plums only if brown rot has previously been a problem.
2. Some products carry warnings about application within certain times of oil sprays. Check labels for directions.

Stage: 3 to 4 weeks before harvest and again 1 to 2 weeks before harvest.

Chemicals: benomyl, carbendazim, iprodione, procymidone, thiophanate-methyl, triforine or vinclozolin.

1. Benomyl, carbendazim, thiophanate-methyl or triforine are preferred particularly in fruit maturing after mid December.
2. It is essential during this period that all diseased fruit is collected and destroyed.

Stage: post-harvest.

Chemicals: treat fruit in a fungicide dip. For details, see 'Post-harvest fungicide dips' below.

Essential if rain prevents spraying close to harvest.

Stage: dormancy.

Within orchard, prune affected twigs, stalks and fruit 'mummies' and burn or bury. This is essential for the control of brown rot.

Post-harvest fungicide dips

Refer to the transit rot information for stone fruit (p. 61).

Dieback

The wood rotting bracket fungi *Trametes versicolor* (*'Polystictus'*), *Schizophyllum commune* (yellow heart rot), *Trametes velutina, Pycnoporus coccineus* and *Stereum strigoso-zonatum.*

Symptoms

At first, a browning and 'papering' of the outer bark of affected branches. These rapidly lose vigour and die. Fungal brackets (fruiting bodies) develop on the diseased wood. Some characteristics of the brackets are given below.

Trametes versicolor: up to 30 mm across, with various coloured bands; smooth surface on the upper side and a honeycomb appearance on the underside.

Schizophyllum commune: soft, whitish, with ragged edges and mushroom-like gills on the underside.

Trametes velutina: similar to *Trametes versicolor* except that the upper side is hairy and the bands are grey to brown.

Pycnoporus coccineus: up to 70 mm across, bright orange-red in colour with a honeycomb appearance on the underside.

Stereum strigoso-zonatum: small, leathery, grey to brown in colour, with a polished black underside.

Source of infection and spread

The brackets produced on old branches and stumps of native and deciduous fruit trees contain large numbers of spores. These are released and carried by wind to nearby stone fruit trees where germination occurs when conditions are suitable. Entry is usually gained through damaged tissues such as pruning cuts and natural wounds. Once established, the fungus grows down through the heartwood causing dieback as it progresses.

The condition is more common in mature trees which have suffered from drought, hail, sunscald, mechanical root damage, overcropping, poor nutrition or severe winter pruning.

Trametes versicolor and *Schizophyllum commune* are the most serious fungi, the others being relatively weak parasites generally found attacking declining trees.

Importance

Generally serious only where trees have been badly stressed by severe pruning, drought or deep cultivation.

Control

1. Prune trees carefully and shape young trees to avoid the necessity for large pruning cuts.
2. Treat cuts with the recommended wound dressing to promote rapid healing.

Wound dressings

After pruning out affected branches at a point at least 200 mm below any visible discolouration, paint the cut surfaces with an anti-mould plastic household paint. Wash the pruning saw blade with methylated spirits after each pruning cut. With healthy trees, prune minimally to shape trees. When large pruning cuts are necessary, they should be done in early summer rather than winter.

Freckle and scab

Venturia carpophila (conidial state: *Fusicladium carpophilum*).

Symptoms

Leaves: poorly defined pale green or brown spots up to 5 mm wide on the underside.

Fruit: at first, circular olive green spots up to 2 mm wide with poorly defined margins. Spots are more common on the exposed side of the fruit. A cork layer develops below the spots and, as the fruit develops, it may become distorted and crack.

Source of infection and spread

The fungus overwinters on the twigs. In spring during wet weather, it produces large numbers of spores which are spread by splashing rain to leaves and fruit. Symptoms may not appear for up to two months after infection.

Importance

Minor in well-sprayed orchards. The disease affects peaches and apricots.

Control

Spray with recommended fungicides.

Stage: three weeks after petal-fall (shuck fall in apricots).

Chemicals: dithianon, mancozeb, thiram, wettable sulphur.

1. Apply three sprays at three-weekly intervals.
2. Sulphur may cause leaf scorch at temperatures above 24°C.
3. Apply copper fungicide at 50% leaf-fall, again at bud movement and 7 to 10 days later.

Leaf curl

Taphrina deformans

Symptoms

Leaves: affected leaves are thickened and distorted with large convoluted blisters which are frequently coloured pink or red.

Fruit: reddish blotches often with surface cracking. Sometimes affected areas are depressed and fruit become misshapen.

Source of infection and spread

Spores of the fungus survive the winter on the surface of trees or on the ground. In the spring, spores are blown into developing leaf and flower buds where infection occurs during cool wet weather.

Importance

Minor in well-sprayed orchards. Affects mainly peaches and nectarines and occasionally apricots.

Control

Spray with recommended fungicides just before budswell.

Stage: early bud movement.

Chemicals: copper oxychloride, lime sulphur.

1. These sprays are essential for leaf curl control.
2. Lime sulphur does not control bacterial diseases. It is preferred where these are not a problem.

Rust

Tranzschelia discolor

Symptoms

Leaves: small, at first pale yellow and later brown spots, covered on the underside by a brown powdery mass of spores. If spots are numerous, large areas of the leaf may be killed.

Twigs: dark brown to black spots breaking open to reveal a light brown mass of spores.

Fruit: tan spots up to 5 mm wide, turning orange. Brown spores may sometimes be present.

Source of infection and spread

The fungus overwinters on twigs and dead leaves. The large numbers of spores produced are spread by wind and rain. Warm wet weather favours infection and disease development. The disease develops from midsummer onwards becoming a serious leaf disease after harvest.

Importance

Minor in well-sprayed orchards.

Control

Spray with recommended fungicides.

Stage: three weeks after petal-fall (shuck fall in apricots).

Peaches and nectarines. Chemicals: dithianon, mancozeb, thiram, wettable sulphur.

1. Apply three sprays at three-weekly intervals.
2. Sulphur may cause leaf scorch at temperatures above 24°C.

Plums. Chemicals: dithianon, thiram, wettable sulphur.

1. Apply three sprays at three-weekly intervals.
2. Do not use dithianon in combination with liquid insecticide formulations on Wilson or Santa Rosa cultivars as ringspotting of fruit may occur.
3. Sulphur may cause leaf scorch at temperatures above 24°C.

Apricots. Chemicals: mancozeb, thiram.

Apply three sprays at three weekly intervals.

Stage: post-harvest.

Peaches and nectarines. Chemicals: dithianon, mancozeb, thiram.

1. This is an important foliage spray and helps to determine next year's crop.
2. Note that apricots need not be sprayed.

Plums. Chemicals: dithianon, wettable sulphur.

As for peaches and nectarines.

Note: Sulphur may cause leaf scorch at temperatures above 24°C.

Shot-hole

Stigmina carpophila

Symptoms

Leaves: small circular tan spots with dark brown margins falling out to give a shot-hole appearance. Where many spots occur together, large ragged holes result. Shot-hole leaf spots caused by this fungus differ from those caused by the bacterium *Xanthomonas pruni.* With bacterial spot, remnants of dead tissue invariably adhere to the edge of the holes.

Fruit: small purplish spots becoming raised and rough with age.

Source of infection and spread

The fungus overwinters on twigs. In spring during wet weather, large numbers of spores are produced and splashed to leaves and fruit. Surfaces must remain wet for several hours for infection to occur.

Importance

Moderate requiring thorough timely sprays.

Control

Spray with recommended fungicides.

Stage: early bud movement.

Chemicals: copper oxychloride, lime sulphur.

Stage: three weeks after petal-fall (shuck fall in apricots).

Chemicals: dithianon, thiram, wettable sulphur.

1. Apply three sprays at three-weekly intervals.
2. Do not use dithianon in combination with liquid insecticide formulations on Wilson or Santa Rosa cultivars as ringspotting of fruit may occur.
3. Sulphur may cause leaf scorch at temperatures above 24°C.

Apricots. Chemicals: mancozeb, thiram.

Apply three sprays at three-weekly intervals.

Stage: post-harvest.

Peaches and nectarines. Chemicals: dithianon, mancozeb, thiram.

1. This is an important foliage spray and helps to determine next year's crop.
2. Note that apricots need not be sprayed.

Stage: 30 to 50% leaf-fall.

Chemical: copper oxychloride.

Lime sulphur 20% (5 L/100 L) is preferred for control of shot-hole if bacterial diseases are not a problem.

Trunk canker

Phytophthora cinnamomi

Symptoms

Leaves: yellowing, wilting and premature fall.

Trunk: long wide sunken areas extending from just below ground level to the first branches and involving up to half the circumference of the trunk. Beneath the sunken bark, an orange-brown discolouration develops in the wood. Copious orange-red gum may exude from the edge of active cankers.

Roots: white feeder roots turn brown and rot.

Source of infection and spread

The fungus causing trunk canker and gummosis is a water mould which thrives in wet soils. The fungus is widespread in coastal Queensland. It is introduced into clean soil with diseased nursery stock and with water and soil movement. The fungus attacks the crown of young trees and, under favourable conditions, invades the trunk and lower branches. As the disease becomes severe the roots are also affected.

Importance

A minor disease affecting stone fruit on peach rootstocks.

Control

1. Plant disease-free nursery trees.
2. Grow in well-drained soils.
3. When a canker appears, cut away all discoloured wood and apply a slurry of a copper fungicide.

Verticillium wilt

Verticillium dahliae

Symptoms

Leaves: slight chlorosis and a dull leathery appearance. Affected trees gradually defoliate with leaf-fall starting at the base of each branch. A few leaves remain at the tip of a branch.

Branches and twigs: affected wood has a brown discolouration in contrast to the normal white wood.

Source of infection and spread

The fungus is a common soil inhabitant and may survive for many years. It invades the roots and grows into the water-conducting tissues of the trunk and branches.

Many crop plants and weeds are susceptible and areas used for growing tomatoes, potatoes and strawberries are commonly affected.

Importance

Uncommon.

Control

1. Avoid growing susceptible vegetable crops such as tomatoes before planting trees.
2. Avoid intercropping with susceptible vegetable crops.
3. Control weeds between trees.

Nematodes

Root-knot nematode
Meloidogyne javanica

Root-lesion nematode
Pratylenchus penetrans

Symptoms

Root-knot nematodes produce galls or swellings on the roots while root- lesion nematodes cause premature death of fine roots. The effects of nematodes cause premature death of fine roots, which reduces growth, particularly in peach and nectarine trees in replant land.

Importance

Can be serious where preplant fumigation is not practised.

Control

1. Fumigate replant sites with the recommended nematicides.
2. Use nematode-resistant rootstocks.
3. Do not replant too soon after removing old trees. Allow the area to revert to weeds or plant recommended cover crops. Do not plant cowpeas, New Zealand blue lupins or rye.

Nematode-resistant rootstocks

The peach rootstock Nemaguard, which is resistant to root-knot nematodes, has performed well with all peach cultivars tested.

Fumigation

1. Replanting individual sites in established orchards. First remove tree roots and cultivate to a depth of 300 mm. Treat a 3.6 m^2 at each site with methyl bromide (one 680 g can) under a 4 m^2 plastic sheet with edges embedded to a depth of 150 mm. Leave the sheet in place for two days.
2. Replanting old orchard land. Remove trees and spell the area for as long as practicable. Allow the area to revert to weeds or cover crop with Sudan grass or maize in the summer and Saia oats, Priebe prairie grass or Wimmera rye grass in the winter. Do not grow cowpeas, rye or New Zealand blue lupins.

Before replanting, prepare soil by ploughing deeply, ripping and discing. Fumigate with EDB 193 at 110 kg/ha.

Viruses

Glengarry spot
Prunus necrotic ringspot virus

Symptoms

On the apricot cultivar Glengarry, sunken red to brown areas develop either as spots or rings on ripening fruit. Symptoms intensify during ripening.

A brown stain in the flesh beneath affected areas may extend to the stone.

Expression of symptoms varies from one season to another.

Source of infection and spread

The disease is commonly associated with the use of infected budwood.

Slow spread of the virus in the field during insect pollination may also occur.

Importance

Not a problem if virus-tested propagating material is used.

Control

Remove affected trees. Use budwood from a proven virus-free source and propagate on virus-free rootstocks. Plant disease-free trees as a block to avoid field spread.

Line pattern
Prunus necrotic ringspot virus

Symptoms

On leaves, yellow to almost white or pale green lines which frequently form 'oak-leaf' and ringspot patterns. The longer lobes of the pattern are usually associated with the main leaf veins. Patterns may appear on leaves of only one growth flush during the season.

Source of infection and spread

The disease is most commonly asssociated with the use of infected budwood, but it may be transmitted through the seed of the rootstock, in clonal rootstocks and even slowly from one mature tree to another by insect pollination.

Importance

Not a problem if virus-tested propagating material is used.

Control

Remove affected trees and replace with virus-free scion material propagated on either virus-free rootstock clones or seedlings from virus-free trees.

Rosette
A virus complex – prune dwarf + Prunus necrotic ringspot

Note on cause

A virus complex usually of prune dwarf virus and Prunus necrotic ringspot virus. Affects peaches and nectarines.

Symptoms

Internodes of affected plants are shortened giving the new growth a bunched compact appearance. In severe cases, leaves may be small and rolled. However, in mild cases, the almost normal summer growth may obscure the spring symptoms. Yield is reduced and susceptible peach cultivars such as Golden Queen may decline.

Source of infection and spread

Both components of the disease may be carried in seed or budwood.

Trees may be infected for some time before expressing symptoms and budwood taken during this time will produce diseased trees. Although only one virus component of the disease may be present in young trees when propagated, the rosette and decline may still appear if the other virus component is introduced in the field during insect pollination. However, this spread is usually slow.

Importance

Not a problem if virus-tested propagating material is used.

Control

Remove affected trees. Use only virus-free scion material and propagate on either virus-free rootstock clones or seedlings obtained from virus-free trees.

POST-HARVEST

Fungi

Transit soft rot

Rhizopus stolonifer

Symptoms

Fruit first show round tan-coloured areas which develop very rapidly into soft wet rot covered by a coarse white fungal growth in which small, white then black, stalked fruiting bodies are produced. The fungus spreads rapidly to other fruit by contact, producing 'nests' of rotted fruit.

Source of infection and spread

The disease is mainly a post-harvest problem and is only occasionally seen in the field. The fungus is common on decaying plant material but a common source of infection is old diseased fruit around the packing shed. Infection generally occurs through injuries or bruises particularly when the fruit is moist. Once established, rapid spread may occur from fruit-to-fruit contact. Warm conditions favour disease development.

Importance

Not a problem where fruit are dipped and hygiene measures followed.

Control

1. Use the recommended post-harvest fungicide treatment.
2. Handle fruit carefully to avoid skin damage.
3. Remove diseased fruit from the packing shed.

Stage: post- harvest.

Chemicals: treat fruit in a fungicide dip.

Essential if rain prevents spraying close to harvest.

Post-harvest fungicide dips (for control of post-harvest brown rot and transit rot)

As soon as practicable after harvesting, dip fruit for 60 seconds in **one** of the following:

a mixture of 50 g benomyl (500 g/kg) and 100 g dicloran (500 g/kg) in 100 L of water plus wetting agent;

a mixture of 50 g carbendazim (500 g/kg) and 100 g dicloran (500 g/kg) in 100 L of water plus wetting agent;

a mixture of 100 ml triforine (190 g/kg) and 100 g dicloran (500 g/kg) in 100 L of water; or

100 g iprodione (500 g/kg) in 100 L of water.

Addition of a wetting agent at a rate of 6.5 mL/100 L dip is desirable for fruit such as nectarines which are difficult to wet.

Note:

1. Any of the dips may be used for peaches, nectarines and apricots. Plums generally need not be dipped.
2. Dipping is essential for long distance travel of fresh fruit and is vital if rain prevents spraying close to harvest.
3. Benomyl, carbendazim and triforine are effective against brown rot only and dicloran is effective against transit rot only. Hence the recommended mixtures.
4. Allow fruit to lose field heat before dipping.
5. The required amount of a powder fungicide should be made into a slurry before adding to the dipping tank. Do not add the powder directly to the tank.
6. Continuous agitation of the dip is required before and during dipping to keep powder fungicides in suspension.
7. After dipping, dry fruit before packing and storage.
8. Topping-up:

   benomyl-carbendazim and triforine-dicloran dips: top-up with a suspension or solution equivalent in strength to that of the original dip.

   iprodione dip: top-up with a suspension double the strength of the original dip.
9. Dips should remain effective for up to two weeks if strained and topped-up with fresh stocks. However, it is advisable to make up a fresh dip every seven days.

Warning: Damage to fruit, especially red pigmented cultivars, can occur if dipped when hot.

Disinfectant for packing and grading equipment

Wash regularly with a hypochlorite-based bleach solution used at household rates, for example 150 mL of 3.5% sodium hypochlorite/L water. As this corrodes some materials, wash thoroughly with clean water after treatment.

Strawberries

FIELD DISEASES

Bacteria and mycoplasmas

Lethal yellows/Green petal
Mycoplasma-like organism

Symptoms

Lethal yellows: affected plants show some inward rolling and purpling of older leaves which flatten to the ground. Young leaves develop yellow margins and remain small with short stalks. Plants become stunted and usually die. Runners from affected plants are similarly stunted.

Green petal: affected plants produce flowers where green leaf-like structures replace the normal petals.

Source of infection and spread

The main source of infection is affected weeds around the crop. The disease is particularly serious when irrigated runner beds attract insects from surrounding drier vegetation in dry summers. The organism is probably spread by a leafhopper. After infection, symptoms may not develop for four to six weeks.

Importance

Periodic epidemics - often serious (lethal yellows). Minor (green petal).

Control

Rogue affected plants and attached runners in runner beds.

Fungi

Armillaria crown rot
Armillaria sp.

Symptoms

Affected plants turn reddish-brown, collapse and die. A cream-coloured fungal growth, sometimes in fans, will be seen beneath the surface of crowns and large roots. Black threads of the fungus resembling shoestrings may also be seen on affected tissues.

Source of infection and spread

The fungus may survive in the soil for many years on roots and stumps of trees. The disease is generally a problem only where strawberries are grown on land which has been recently cleared. Damp soil situations favour the development of the fungus.

Importance

Rare.

Control

Remove all tree stumps and large roots before planting strawberries on land which has been recently cleared.

Black spot
Colletotrichum acutatum

Symptoms

Fruit: firm, circular, dark brown, sunken spots up to 20 mm wide. White fungal growth and pink spore masses develop on the spots during moist weather.

Leaf and flower stalks: dark slightly depressed elongated spots. Found occasionally on plants grown under harsh conditions.

Source of infection and spread

The fungus infects leaves and green fruit and, after penetration, ceases growth and remains dormant in the tissues. It resumes growth in fruit when they begin to ripen and in leaves when they die. The spore masses produced on old undecomposed leaves are the main source of infection. The spores are spread by splashing water and on the hands of pickers. The fungus may also be introduced in diseased runners. Fruit showing no symptoms at harvest may develop black spot during transit and storage.

Importance

Can be serious if spray programme not followed.

Control

Spray with recommended fungicides.

Fungicide sprays

Stage: flowering until final harvest.

1. For control of grey mould and black spot spray with dichlofluanid every 7 to 14 days. Use the shorter interval during wet weather. Do not use a wetting agent. Withholding period is one day.
2. Add benomyl for the control of powdery mildew. This disease is generally not a problem on cv. Redlands Crimson.
3. If grey mould is a problem, despite the use of dichlofluanid, add iprodione or vinclozolin.

Stage: when plants are not in production.

(in runner production beds, ratoon crops and new plantings)

Spray with benomyl every 14 to 28 days. Use the shorter interval when prolonged wet weather occurs.

Note: Do not use benomyl after releasing predatory mites in an integrated pest management system as the chemical is highly toxic to the mite.

Crown and fruit rot

Phytophthora nicotianae var. *parasitica*

Symptoms

Plants: at first the outer then the inner leaves collapse, turn brown and die. The crown tissues are discoloured, and, if the disease is severe, roots are black and rotted.

Fruit: tan-coloured spots enlarging to eventually cover the berry.

Bleaching of the skin may occur on mature fruit. If affected fruit are cut lengthwise, a brown discolouration of the water-conducting tissues beneath diseased areas will be seen. In moist weather, a white fluffy fungal growth may cover entire clusters of fruit.

Source of infection and spread

The fungus is a soil-inhabiting water-mould which requires water for spore production and plant infection. Consequently, the disease is most serious during prolonged wet weather and in low lying or wet areas. Fruit in contact with the soil are most susceptible to fruit rot but spores may be splashed from the soil to fruit on polythene mulch. Crown rot losses are most common in late spring or summer.

If diseased fruit are packed when wet, the fungus may be spread rapidly causing nests of rotted fruit.

Importance

Widespread in very wet seasons but generally minor.

Control

Plant strawberries only in well-drained soils on raised beds covered with polythene mulch.

Preplant fumigation

Apply, at label rates, methyl bromide + chloropicrin mixture or dazomet.

Eye spot

Mycosphaerella fragariae

Symptoms

Leaves: small, reddish to purple spots, the centres of which dry out to a light grey or white colour. Small black fruiting bodies of the fungus may appear in the centre of older spots. When numerous, spots may cause a rapid scorching of the leaves and a gradual defoliation of the plant.

Leaf and fruit stalks, stolons and and calyces may show similar spots.

Fruit: firm dark brown sunken spots up to 3 mm wide occasionally occur on fruit.

Source of infection and spread

The main sources of infection are diseased ratoon crops or undecomposed strawberry residues in the soil. The fungus may also be introduced on diseased runners. Spores produced in the small black fruiting bodies are spread during wet weather. Cool conditions favour disease development. The disease is worst in low lying areas or where plants are shaded.

Importance

Minor in well-sprayed crops.

Control

1. Reject severely affected runners and remove diseased leaves from others before planting.
2. Do not plant strawberries in low-lying or shady areas.

Fusarium wilt

Fusarium oxysporum f.sp. *fragariae*

Symptoms

In warm weather, the outer leaves and then the whole plant wilts suddenly and dies. However, in cool weather, some of the central leaves remain healthy with others slightly rolled and yellow. When the crown of an affected plant is cut lengthwise, a brown discolouration of the water-conducting tissues often extending into the leaf bases will be seen. In advanced stages the crown rots completely.

Source of infection and spread

The fungus is a soil inhabitant and may survive there for long periods.

It enters the plant through the roots and grows into the water-conducting tissues of the crown and leaves. It is most serious in summer.

If an affected plant produces runners, the fungus may spread through the stolon from the parent plant into the runner. The use of these for planting is frequently responsible for the spread of the disease.

Importance

Serious in some areas in very warm seasons.

Control

1. Use runners from disease-free areas.
2. Either fumigate with methyl bromide + chloropicrin at label rates, or do not plant infested areas.

Grey mould

Botrytis cinerea

Symptoms

On fruit, light brown spots with indefinite margins increasing rapidly in size until the whole fruit becomes brown and rotten. During moist weather, a grey powdery growth develops on affected areas. With time, berries tend to dry out and harden.

Source of infection and spread

The fungus is a common inhabitant of dead or dying plant material in the soil. Large numbers of spores are produced and spread by wind and rain. These germinate on dead flowers or leaves and penetrate fruit. Fruit in contact with wet soil are also readily infected. Cool, moist weather favours the disease.

Importance

Can be serious if spray programme not followed.

Control

1. Spray with recommended fungicides.
2. Plant on polythene mulch to prevent fruit contacting the soil.

Fungicide sprays

Refer to the black spot (p. 63).

Leaf blight

Dendrophoma obscurans

Symptoms

On leaves, large spots up to 30 mm wide, at first reddish-purple in colour but later developing a dark brown centre surrounded by a light red margin.

Spots may be circular to elongate in shape, with concentric markings or V-shaped areas extending in from the leaf margin. Small black fruiting bodies of the fungus form on affected areas.

Source of infection and spread

The main sources of infection are diseased ratoon crops or undecomposed strawberry residues in the soil. The fungus may also be introduced on diseased runners. Spores produced in the small black fruiting bodies are spread during wet weather. Warm conditions favour disease development, and it is commonly a problem in runner beds during summer. The disease is worst in low-lying areas or where plants are shaded.

Importance

Minor in well-sprayed crops.

Control

1. Reject severely affected runners and remove diseased leaves from others before planting.
2. Do not plant strawberries in low-lying or shady areas.

Leak

Rhizopus arrhizus

Symptoms

A soft watery rot of ripe fruit, generally starting at injuries.

Under moist conditions, affected fruit become covered with a mass of white fungal growth which eventually produces large numbers of small black spore bodies on stalks.

Fruit eventually become black, hard and shrivelled.

In boxes and punnets, large amounts of liquid may ooze from affected fruit, hence the common name, 'leak'.

Source of infection and spread

The fungus is common on dead or dying plant material in the soil. Large numbers of small black spores are produced and spread by wind. Infection occurs mainly through injuries to the skin and, consequently, the disease is often serious after heavy storms or when fruit have been injured by insects. Hot wet weather favours the development of the fungus. It attacks only ripe or over-ripe fruit and is mainly a post-harvest problem. However, it may be common in the field when over-ripe fruit is not harvested.

Importance

Common and may be serious late in the season particularly in over-ripe fruit.

Control

1. Remove over-ripe and affected fruit at each harvest.
2. Remove and bury waste fruit from around the packing shed.
3. Do not pack over-ripe or damaged fruit.
4. Handle fruit carefully during harvesting and packing.
5. Cool fruit promptly after harvesting.

Leather rot

Rhizoctonia solani

Symptoms

A hard brown rot of berries starting on the side of the fruit touching the soil. Soil particles generally adhere to the affected area.

Source of infection and spread

The fungus is a common soil inhabitant and infects fruit in contact with soil.

Importance

Rare since widespread use of plastic mulch.

Control

Not warranted.

Powdery mildew

Sphaerotheca macularis

Symptoms

Leaves: a white, powdery coating generally on the lower surface and on the stalks. Leaves of some cultivars may show a reddish colour on the surface.

Fruit: a white powdery coating on fruit, fruit stalks and calyces.

The surfaces of the fruit are hardened and, with ripening, generally crack.

This is often confined to the stem-end of the fruit.

Source of infection and spread

The fungus survives on diseased ratoon crops. The white mildew contains large numbers of spores which may be carried long distances by wind. High humidity and heavy shade favour infection but wet conditions are not necessary.

Importance

Minor in well sprayed crops.

Control

Spray with recommended fungicides.

Fungicide sprays

Refer to the black spot entry.

Sclerotinia crown rot

Sclerotinia sclerotiorum

Symptoms

The bases of the leaf stalks of the older outer leaves are killed and the leaves collapse. In wet weather, white fungal growth will be found on the affected tissues and later, large black irregular resting bodies (sclerotia) form within the crown. Fruit may also be affected.

General

For a description of the life cycle of the fungus, see Sclerotinia fruit rot (p. 66).

Importance

Minor.

Control

Avoid planting in heavily infested areas.

Sclerotinia fruit rot

Sclerotinia sclerotiorum

Symptoms

A soft watery rot of fruit which, in moist weather, becomes covered with a white fungal growth in which large black irregular resting bodies (sclerotia) form.

Source of infection and spread

The sclerotia enable the fungus to survive in the soil for long periods. In cool moist weather, those near the surface germinate to produce at soil level small cream mushroom-like bodies called apothecia. These contain large numbers of ascospores which are forcibly discharged and carried by wind. Before the fungus can attack healthy fruit, it must first germinate on and infect dying plant material such as old petals and leaves. It then invades those fruit in contact. Cool wet weather favours infection but fogs, mists, dews and irrigation provide enough moisture for infection. Fruit on plants with a dense growth of leaves are more likely to be affected.

Many cultivated plants are suspectible to Sclerotinia rot.

Importance

Minor.

Control

Generally not warranted.

Sclerotium crown rot

Sclerotium rolfsii

Symptoms

A sudden wilt and collapse of plants. If the crowns of affected plants are examined, a white fungal growth will usually be seen on the leaf bases.

Small spherical, white then later brown to black, resting bodies (sclerotia) form in the fungal growth.

Source of infection and spread

The fungus survives in the soil as the small dark sclerotia and is active in decaying plant residues. It is most serious in summer, particularly in ratoon crops and runner beds.

Importance

Generally minor. Found more frequently in ratoon crops.

Control

1. Prepare land thoroughly before planting to ensure that crop residues have thoroughly decomposed.
2. Remove dead leaves from around bushes.

Scorch

Diplocarpon earlianum

Symptoms

On leaves, small purplish irregular spots enlarging with age and the centres becoming brown. Small raised glistening fruiting structures of the fungus appear on the upper surface. Spots may coalesce causing the leaves to brown, dry out and appear scorched. Leaf and flower stalks may be girdled by spots.

Source of infection and spread

The main sources of infection are diseased ratoon crops or undecomposed strawberry residues in the soil. The fungus may also be introduced on diseased runners. Spores produced on the spots are spread during wet weather.

The disease is worst in low lying areas or where plants are shaded.

Importance

Minor in well sprayed crops.

Control

1. Reject severely affected runners and remove diseased leaves from others before planting.
2. Do not plant strawberries in low-lying or shady areas.

Verticillium wilt

Verticillium dahliae

Symptoms

Cultivars vary in their expression of this disease. The most common symptom is a slow wilting and death of the older leaves. Often only one or two leaves will be wilted at any one time and these generally die before others begin to wilt. This eventually results in an outer ring of dead leaves with a central tuft of green erect leaves. These may also succumb and the plant finally dies.

Source of infection and spread

The fungus is a common soil inhabitant and may survive there for long periods. It may also be introduced in diseased runners. It invades the plant through the roots and grows into the water-conducting tissues of the crown.

The disease is generally serious only in the cooler months of the year and, in Queensland, plants which have not died by mid-September often recover.

The fungus attacks many other crop plants and weeds.

Importance

Often not noticed as plants grow out of the symptoms. Sometimes serious.

Control

1. Plant approved runners.
2. Do not plant in infested areas.

Preplant fumigation

Apply, at label rates, methyl bromide + chloropicrin mixture or dazomet.

Nematodes

Root-knot nematode

Mcloidogync hapla

Symptoms

Small galls or swellings on the secondary roots with the main roots short and ending in distinct swellings. The root system is reduced and densely matted. Heavily infested plants are stunted and yield poorly.

Source of infection and spread

The nematode is a soil inhabitant and has a wide host range including cucumber, peanut, potato, tomato and capsicum. Larvae enter roots causing the formation of galls. Nematodes can be spread by planting infested runners and in soil washed down slopes or adhering to farm machinery.

Importance

Causes unthriftiness in some fields

Control

1. Plant approved runners.
2. Apply preplant nematicides.

Root-lesion nematodes

Pratylenchus spp.

Symptoms

Premature browning and death of roots. Infested plants may grow satisfactorily until mid season, after which they decline and are unsuitable for ratooning.

Source of infection and spread

The nematodes are soil inhabitants and other hosts include cowpeas, French beans, lab lab, velvet beans and mung beans. Spread occurs in infested runners and in contaminated soil.

Importance

Causes unthriftiness in some fields

Control

1. Plant approved runners.
2. Apply preplant nematicides.

Preplant nematicide treatment

Necessary only in areas infested with the strawberry root-knot or root-lesion nematode. Treat soil with ethylene dibromide (EDB). Fumigate at least two weeks before planting at recommended label rates. Otherwise treat the soil one to seven days before planting with fenamiphos spray applied as an overall treatment at the rate of 24 L/150 to 300 L water/ha. Incorporate to a depth of 100 mm with a rotary hoe or offset disc as soon as possible after application.

Apply while the soil is moist after irrigation or rain.

Where control of fungi, weeds and insects is also required, use either methyl bromide + chloropicrin mixture or dazomet. Refer to labels for details.

Strawberry bud nematode

Aphelenchoides besseyi

Symptoms

Affected plants are smaller with a more compact appearance and the inner leaves are narrower, more crinkled and darker in colour than normal. However, symptoms are generally not obvious after early August in Queensland, but reappear on the runners in November-December. Affected plants produce fruit later in the season than healthy plants and a proportion of fruit is small and deformed.

Source of infection and spread

The nematode is usually introduced in infested runners. It is also common in the seedheads of a number of grasses.

In the strawberry plant, the nematode lives in the crown between the small developing leaves and flower buds but rarely enters the tissues. The malformation of leaves and flowers results from the nematodes feeding on the growing point.

Importance

Minor.

Control

1. Plant runners free from bud nematode.
2. In runner production areas, remove affected plants as soon as they are noticed.
3. Mow grasses around crop areas.
4. Rogue affected plants or treat with the recommended nematicide.

Nematicide treatment

Inspect the crop within one month of planting and with gloved hands and applicator, place 1 g fenamiphos (100 g/kg product) granules in the heart of each affected plant. Water heavily after treatment.

Viruses

Mild yellow edge virus combined with other viruses

Virus diseases have not been a problem in Queensland for many years because of the operation of the Strawberry Runner Approval Scheme. However, yellow edge disease, caused by a combination of other viruses with mild yellow edge virus, was formerly common and resulted in stunted plants with small pale yellow fringed leaves.

S. R. Hampson, Government Printer, Queensland—1989

P-013